Foundations in Earth Science

Adventures in Nature and Mathematics

Carl N. Drummond

Department of Geosciences
Indiana University Purdue University Fort Wayne

KENDALL/HUNT PUBLISHING COMPANY
4050 Westmark Drive Dubuque, Iowa 52002

ISBN 0-7872-5154-2

Printed in the United States of America
10 9 8 7 6 5 4 3 2

Contents in Brief

Preface

I've heard the comments for years, as has every other mathematics professor or teacher. "My students can't do elementary mathematics." "Why do my students do such outrageous things?" "Don't you teach exponents?" Frustrating comments, as we know we've taught the topics, think we have done a reasonable job, and, in many cases, would swear that our students have "learned" what we taught them. Why won't the science professors and teachers just leave us alone?

Well, it turns out that helping students learn how to be quantitative thinkers is incredibly difficult. Analyzing data is not a simple task. From the accusations of Disreaeli, to statistical "paradoxes" that turn out to be all too common, to the public's distrust of scientific "conclusions", we have ample evidence that a greater understanding of analyzing data and drawing conclusions from the analyses is essential if we are to ever achieve the goal of a scientific and mathematically literate populace.

The National Council of Teachers of Mathematics, in *its Curriculum and Evaluation Standards for School Mathematics* (1989) has established the following four standards as crucial to the development of mathematical power: mathematical problem solving, mathematical communication, mathematical reasoning, and the establishment of mathematical connections. In the reformation of school mathematics, these four principles become both end products and the day-in day-out cornerstones of instruction and student mathematical activity. This reformation assumes that mathematics is something for students to "make sense of" rather than something to memorize, and that a major reason to learn mathematics is so that we can make sense of data and the world around us. Analysis of data, communication of solution processes, and defending one's conclusions become central processes in mathematics classrooms and instruction.

This shift in instructional thinking as to how to teach mathematics will be of major benefit to science, and those who's task it is to teach science. In itself, however, it will not be enough. All the evidence points to the fact that for students to really learn and understand, that teaching and instruction must be done in context. Simply put, the paradigm of learning how and then applying the how to different situations is not sufficient for the development of quantitative thought.

Foundations in Earth Science is a serious step forward in the evolution of introductory science instruction. The power of the study of data analysis in the context of learning beginning Earth science is one that transcends scientific disciplines. It is especially appropriate as professors retool introductory courses sot hat they serve the twin goals of general education and introductory courses for science majors. Work your way through *Foundations in Earth Science*, as I have, and you will see a serious support text which can improve the quantitative reasoning abilities of both science and non-science majors. For too long we have assumed that science majors can analyze data naturally, and that non-science majors can not analyze data at all. *Foundations in Earth Science* will improve the quantitative reasoning skills of all students who use it, and I regret that I did not have the opportunity to study introductory science in this manner. I do, however, look forward to my children and my students becoming powerful quantitative thinkers as they study and learn Earth science and data analysis with *Foundations in Earth Science* and the generations of texts which it will inspire.

Jim Hersberger, Ph.D.
Professor of Mathematics Education

Detailed Contents

Forward

Science represents a highly formalized way of approaching nature. It is based upon the multifaceted concepts of logic and reason, rationality and testing, replication and falsification. However, the functional foundation of science is its integration with mathematics. Mathematics is in essence the universal language of science. It allows us to speak exactly and plainly about the world around us. It provides a logical mechanism by which our concepts of nature can be tested and refined. Through it models of the world around us can be constructed such that our collective understanding not only of what nature is like, but how it works, can grow and flourish. Modern scientific inquiry rests solidly upon the foundation of mathematics.

It is the purpose of this book to provide students of introductory science with a step-by-step guide to exploring and understanding the mathematical basis of science. Through the use of this book, students will achieve a more complete knowledge of the concepts presented in their lectures, readings, and laboratory assignments. By allowing students to enlarge their understanding of the mathematics common to the Earth sciences, this book frees them to think more carefully and more creatively about the material they are learning.

Foundations in Earth Science is designed to serve as a supplemental text to introductory lecture sections in Geology, Geography, Environmental Geology, Oceanography, and Astronomy, although the concepts presented here are applicable to any introductory science class. Numerous examples and exercises have been developed which touch upon the various geological subject areas. However, any course in the Earth sciences can readily benefit from the inclusion of material presented herein. The ten chapters of the book are broken into four general groupings:

Chapters 1, 2 and 3 concern the concepts of magnitude, uncertainty, and functional relationships.

Chapters 4 and 5 provide introduction to the construction and interpretation of graphical data.

Chapters 6 and 7 serve as introductions to the analytical tools of statistics and logarithmic relations.

Chapters 8, 9, and 10 describe various ways scientists use mathematical concepts to bring order to the complexity of nature.

The mathematical foundations of science are at times difficult for students to grasp and appreciate. This book presents not only a workbook for students to learn the fundamentals of mathematical science, it also can serve as a useful reference to the range of applied mathematics students will encounter in such diverse fields as the social sciences, economics, business, and education. As such, a foundation in mathematical literacy will serve students well irrespective of their chosen career path.

Carl N. Drummond, Ph.D.
Fort Wayne, Indiana
30 April, 1998

Chapter 1 - Magnitudes: Relative and Absolute

Introduction

What is it that sets science apart from other forms of human creativity? One of the most commonly considered points is that modern science is an extremely quantitative endeavor. Images of scientists hard at work in front of a myriad of computer screens and digital readouts, or diligently wringing one analysis after another from a complex apparatus are common to our general perception of what it is that scientists do. While science can be defined in a number of ways, there is no question that the making of measurements of natural phenomenon is central to the pursuit of scientific knowledge. How is it that scientists measure and describe nature in quantitative terms? In this chapter we will explore the ways in which scientists think of and express the sizes and ages of things in terms of relative and absolute relationships.

The Relative

From the inception of human intelligence we, as a species, have been aware of the concept of the relative. A fundamental aspect of being intelligent is recognition of the magnitude of something in one's environment, and subsequent comparison of that object to other similar objects. Scientists compare previously unknown or unrecognized features of the world to those which are at least partially familiar or understood. This process of comparative analysis starts with the concept of relative magnitude.

Comparisons of relative magnitude and age are central to science.

Not all stones are the same size, nor do all stars shine with equal brightness; nature surrounds us with the complexity of its variety. That variety is expressed not only in the range of *what* it is we observe, but also in the *sizes* and *ages* of those things that might otherwise be considered to be the same. In the sciences two important relative comparisons are commonly made: comparisons of magnitude, and comparisons of age. By magnitude we mean how big, how long, how wide, how deep, how hot or cold something is, by age we mean younger versus older. The Earth is larger than the moon, a mountain larger than the hills that surround it — these things we can observe and understand clearly in terms of relative sizes. Likewise, it is possible to assign relative ages to many features within our environment. A soil is younger than the rock on which it develops, a river's sediment younger than the rocks from which it was eroded.

Comparison of like types.

In order to assure ourselves of an accurate comparison of relative magnitude or age it is first necessary to confirm that a valid comparison is being made. It makes no sense to ask which has the greater magnitude — water at 98° C or a boulder 3 m across. In order for a valid and meaningful comparison to be made, it is necessary that the types of measurements in consideration are of the same type. In addition, it is critically important to describe similar types of measurements in terms of the same units of measure. Without looking it up, which is hotter — a day at 40° C or one at 65° F? In order for correct mathematical calculations to be carried out on observed and recorded data, the units of measure must be internally consistent. Relative comparison and mathematical analysis are most easily conducted when the units are internally consistent. Scientific observations or interpretations are commonly presented in terms of a mathematical relationship or formula. In order for the mathematical calculations to be carried out correctly it is necessary that the measurements and empirical constants within the equation be expressed in the same units. Failure to correctly account for the units within a formula is the basis of many of the mathematical errors students commonly make in science classes. Throughout the semester, you should repeatedly refer to the unit conversion chart found in Appendix I.

Scientific Notation

The modern mathematical system in common usage is based upon a method of counting and numerical representation known as the base 10 system. When we start counting and writing down the numbers counted we proceed in the following way:

0, 1, 2, 3, 4, 5, 6, 7, 8, 9

Each quantity is represented by a different numeral. Five planets in a solar system are different from six by one planet. Each whole number value is expressed by a unique digit, but what happens when we have one more than 9? At that point no new symbol is used to express the notion of the number ten, rather, we combine the digits we already have to create the symbol 10. This notion of place counting is central to our modern mathematical structure, and without it arithmetic would be considerably more complex. Importantly however, you must realize that this base ten system with which we are so familiar is not the only way to express a numerical value. Two methods of keeping track of a number's value are digital (base 2) and hexadecimal (base 16). These different systems are commonly used in a variety of computer applications.

Base 10, 2, and 16 number systems.

One of the most fascinating aspects of the use of mathematics in the sciences is the wide range of very large and very small values encountered. Because scientists are forced to deal with such a range of possible values, a system of shorthand notation has been developed to make the manipulation of such numbers easier. Consider the number

14560000

That is a very large and a bit cumbersome number with which to deal, especially if one is asked to multiply that number by 132.

In order to facilitate such calculations, the concept of scientific notation has been developed. The foundation of this short hand notation is the concept of the exponent. Exponential relationships are a bit daunting at first glance, what does the number

$$17^5$$

Calculating an exponent.

represent? It simply means calculate the product

17 x 17 x 17 x 17 x 17

which is, you will agree, a very large number. The utility of exponential notation is that it allows us to represent very large numbers in a compact fashion. Consider the series of numbers

10, 100, 1000, 10000, 100000, 1000000

What do these numbers have in common? They are the products of multiplying the number 10 with itself over and over again. The concept of scientific notation allows us to represent these and other large numbers in terms of powers of ten. This series of six numbers could be represented as the product of the number 1 multiplied with the following exponential values

Exponential notation in the base 10 system.

$$10^1, 10^2, 10^3, 10^4, 10^5, 10^6$$

to produce the following shorthand expressions

$$1x10^1, 1x10^2, 1x10^3, 1x10^4, 1x10^5, 1x10^6$$

Consider the values 132 and 14500. These are large and slightly awkward numbers with which to deal. However, they can be expressed in a very simple way by applying scientific notation to produce the following

$$1.32x10^2 \text{ and } 1.45x10^4$$

How were these expressions derived? The process is actually quite simple when we remember the idea of place values in our base 10 system of numbers. Ten to the second power means 10 x 10, or 100. One hundred multiplied with 1.32 is 132 — the number with which we started. Likewise 10^4 is 10000, and that number multiplied with 1.45 is 14500.

There is a very straightforward way to find the correct scientific notation for a large number: one simply moves the "decimal place" to the left, keeping track of the number of numerals that are passed over. For instance the number 3482 is expressed in scientific notation as 3.482×10^3. Consider the following relationships:

Scientific notation of numbers larger than 1.

$$3482 \times 1 = 3482 \times 10^0 = 3482$$
$$348.2 \times 10 = 348.2 \times 10^1 = 3482$$
$$34.82 \times 100 = 34.82 \times 10^2 = 3482$$
$$3.482 \times 1000 = 3.482 \times 10^3 = 3482$$

By shifting the decimal place to the left three times, one derives the accepted scientific notation representation of the number 3482. Importantly, the same type of procedure is used to represent very small numbers as well. The decimal representation of the fraction 2 / 1000 is 0.002. How can this very small number be expressed in scientific notation? The process is identical to the one used for large numbers, except the "decimal place" is moved to the right.

Scientific notation of fractional numbers.

$$0.002 \times 1 = 0.002 \times 10^0 = 0.002$$
$$0.02 \times 0.1 = 0.020 \times 10^{-1} = 0.002$$
$$0.2 \times 0.01 = 0.200 \times 10^{-2} = 0.002$$
$$2 \times 0.001 = 2.000 \times 10^{-3} = 0.002$$

The fractions 1/10, 1/100, 1/1000 are represented in exponential notation by the use of a negative exponent: 10^{-1}, 10^{-2}, 10^{-3}. Thus, the fraction 2/1000 is simply 2×10^{-3} in scientific notation.

Numbers raised to the 0 power.

Importantly, there is one special case. How are the counting numbers 0 through 9 expressed in scientific notation? One of the important rules of exponential relationship is that *any* number raised to the zero power is by definition 1. Thus, the number 5 can be represented as 5×10^0. Typically, however the counting numbers are not expressed in scientific notation, and the 10^0 is simply omitted.

Beyond reducing large numbers to smaller expressions, what has been gained? It is not immediately clear what is the point of expressing the number 1675212 in the form 1.675212×10^6. If anything, this number symbol is now larger and more cumbersome than before! The true utility of scientific notation is not fully appreciated until one understands the laws of exponents.

Laws of Exponents

$$x^0 = 1$$
$$x^1 = x$$
$$x^a y^b = (xy)^{a+b}$$
$$x^a/y^b = (x/y)^{a-b}$$
$$x^{1/a} = \sqrt[a]{x}$$
$$(x^a)^b = x^{ab}$$

Table of the laws of exponents.

Thus, the product of the two numbers 132 x 2000 can be easily calculated using scientific notation.

Calculation using scientific notation.

$$(1.32 \times 10^2) \times (2 \times 10^3) = 2.64 \times 10^{2+3} = 2.64 \times 10^5$$

which is the number

264000

Clearly, the application of the laws of exponents allows for rapid calculation of arithmetic expressions involving very large or very small numbers, and that is the true basis of its utility to science.

Ranges in Time and Space

Perhaps the most daunting intellectual challenge facing students of introductory level science is the range of temporal and spatial magnitudes commonly considered by scientists. In order to be fully conversant in the concepts of relative and absolute magnitude it is necessary to appreciate time and space. The following is a brief listing of some of the common lengths and durations considered in the sciences.

Examples of absolute magnitudes expressed in scientific notation.

Distance	***Approximate Length (m)***
Earth to most remote galaxy	4×10^{25}
Earth to nearest large galaxy	2×10^{22}
Earth to nearest star	4×10^{16}
One light-year	9.46×10^{15}
Average Earth-Moon distance	3.8×10^8
Average radius of the Earth	6.4×10^6
Typical altitude of Space Shuttle in orbit	2×10^5
Length of a soccer field	1×10^2
Length of stride of typical adult male	1.7×10^0
Diameter of coarse sand	1×10^{-3}
Diameter of clay grain	4×10^{-7}
Diameter of a hydrogen atom	1×10^{-11}
Diameter of an atomic nucleus	1×10^{-14}

Examples of absolute durations expressed in scientific notation.

Duration	***Approximate Length of Time (s)***
Age of the universe	5×10^{17}
Age of the Earth	1.3×10^{17}
Age of the United States	6.8×10^{9}
Age of an "average" college student	6.3×10^{8}
One year	3.2×10^{7}
One day	8.6×10^{4}
Length of a round of golf	1.44×10^{4}
Time between normal heartbeats	8×10^{-1}
Period of audible sound waves	1×10^{-3}
Period of radio waves	1×10^{-6}
Period of vibration of an atom in a solid	1×10^{-13}

Clearly, nature manifests itself to us in an astonishing range of sizes and durations. Scientific notation is a great help in dealing with these immensities in simple and straightforward ways.

The Metric System

Throughout your exploration of the sciences, you will continually be forced to consider four very important standards of measure: length, mass, time, and temperature. Throughout human history, civilization has recognized the need for standardization of measurements. Whether it be furlongs or fortnights, cubits or centimeters, people have always needed a way to quantify their world and convey to others the information they have gained. At the present, the universally accepted system of measures is known as *Le Systeme International d'Unites*, or in its English translation The International System of Units (SI). This system of measure is nearly identical to the more commonly known metric system. There are seven fundamental units of measure in the SI system, of which the first four are most commonly considered in the Earth sciences.

Basic units of measure in the metric system.

Measure	***Unit***	***Symbol***
Length	Meter	m
Mass	Gram	g
Time	Second	s
Temperature	Kelvin	K
Electrical Current	Ampere	A
Amount of Substance	Mole	mol
Luminous Intensity	Candela	cd

From these fundamental units of measure, it is possible to derive other important units. Area is defined as length multiplied with length, velocity is length divided by time, and so forth. The more familiar metric system has as its set of fundamental units the meter, the gram, the second, and the degree Celsius.

From these bases of measure, it is often convenient to consider subdivisions or multiples of the fundamental unit. Because of the utility of the base 10 system in mathematics, it has also been adopted in the metric system. Scientific notation provides the mechanism by which we can understand these larger and smaller versions of the fundamental units of measure. Typically, such larger multiples or smaller divisions are applied to measures of length, mass, and units derived from those two measures and are denoted by the addition of a prefix to the general unit term. The following are commonly used metric prefixes:

Prefixes of the metric system and their relation to the fundamental unit.

Prefix	***Symbol***	***Factor***
kilo	k	10^3
deci	d	10^{-1}
centi	c	10^{-2}
milli	m	10^{-3}
micro	μ	10^{-6}
nano	n	10^{-9}

Thus, it is possible to understand that a common unit of measure such as the kilometer means $1\text{x}10^3$ meters, or 1000 meters, while a milligram is $1\text{x}10^{-3}$ gram.

In addition to these universally accepted units of measure, there are still in use several other systems. The most prevalent is known as the English system of weights and measures. Common in this system are units of length measure such as the foot and inch, and units of weight measure such as the pound. While it is desirable to record and report all scientific measures in terms of SI units, it is still common to find data presented in these older units. As such, it is necessary to know how to convert from one system to the other. Appendix I provides a summary of common conversions applicable to most calculations in the Earth sciences.

Chapter 1 - Exercises

Name ______________________ Date ______________________

Instructor ______________________ Section ______________________

1) Explore one of the World Wide Web sites provided by your instructor. Record at least 5 examples of measurements of magnitude or duration provided at that site. Be sure that the values you record are in the same units of measure.

Measurement	Units	Description
A		
B		
C		
D		
E		

In the space provided below, write a brief discussion of the range of relative and absolute magnitudes you have recorded above.

2) Convert the following values from decimal notation to scientific notation.

a. 123

b. 0.0004716

c. 450117

d. 0.25

e. 0.000174

3) Convert the following from scientific to decimal notation.

a. 3.475×10^{-6}

b. 2.76×10^{-3}

c. 7.446×10^{7}

d. 1.4462184×10^{-2}

e. 8.0×10^{4}

4) Using the laws of exponents calculate the following, express in scientific notation, and show all calculations.

a. $(1.65 \times 10^{-3})(6.44 \times 10^{5})$

b. (432)(166478)

c. $(7.8 \times 10^{2}) / (4 \times 10^{-8})$

d. $(2.1 \times 10^{2})^{3}$

e. $(5.46 \times 10^{5})(100)$

f. $(4.471 \times 10^{-3}) / (6.4 \times 10^{2})$

g. 26^{-4}

h. $8^{1/3}$

i. $(22)^{3}$

j. $(1.99 \times 10^{2})(7.119 \times 10^{7})$

5) Pick five things you do in normal everyday life, estimate their duration in the units of time in which you would typically measure them. Convert that estimate to seconds. Express the total duration in scientific notation and provide all calculations.

a.

b.

c.

d.

e.

6) Carry out the following conversions:

a. 1.476×10^4 mm = ________ m

b. 4.92×10^{-2} kg = ______ mg

c. 2.000×10^3 s = ______ hrs

d. 6.91×10^{-3} km = ______ cm

e. 6×10^9 g = ______ kg

7) Calculate the following; express the result in scientific notation. Show all work.

a. A landslide is found to cover a rectangular area of 1.47×10^3 m by 1.13×10^2 m, calculate its area.

______ m^2

b. A rising tide was found to have increased local sea level 7.0×10^{-1} m in 6 hours. What was the *average* rate of sea level rise in cm/s?

______ cm/s

c. The volume of a cube of rock is found to be 2.75×10^{1} m^3; what is the length of one side in meters?

________ m

d. If the density of that cube of rock is know to be 3.20 g/cm^3, what is the mass of the cube in kg?

________ kg

8) Using Appendix I, calculate the following conversions; use scientific notation and show all work.

a. 14.75×10^{3} yards =

________ m

b. 25.7 °C =

________ °F

c. 4.27×10^2 lbs =

_________ kg

d. 6.53×10^2 cm =

_________ in

e. 9.658×10^{16} m =

_________ light years

Chapter 2 - Uncertainty: Accuracy and Precision

Introduction

It has been established that science deals with the making of measurements, and, as such, scientific results are often presented or portrayed as nearly infallible. In fact, nothing could be further from the truth. Every scientific experiment, every measurement and observation, has associated with it some uncertainty. Understanding the origins of uncertainty and developing ways to reduce that uncertainty are typical pursuits of a modern scientist. While the details of error analysis and minimization may seem somewhat tedious to the lay person, for many scientists it is exactly that type of problem-solving which drew them to their careers. In this chapter we will explore the forms of uncertainty intrinsic to scientific measurements, and then consider how those uncertainties propagate through a calculation, as well as develop a simple technique for reducing the likelihood of accidentally introducing arithmetic error into our calculations.

The Nature of a Measurement

A scientific measurement is properly expressed only when it contains two components, the measurement's magnitude, and its reliability. Typically when we report a measurement of some feature of our day-to-day lives, we only include the magnitude: to cook a roast the oven is set at a temperature of 350 °F; a car is said to be traveling at 65 miles per hour. Surely these measurements are not exact. Anyone who has had the misfortune to burn dinner or receive a speeding ticket will know that these common measurements have associated with them some range of uncertainty. The oven is not *exactly* 350 °F, the car is not traveling *exactly* 65 miles per hour. Why is this uncertainty, which we know to be there, not commonly discussed or reported except in embarrassment to hungry dinner guests or to the state trooper who just stopped us? The obvious reason is that under normal conditions the uncertainty is sufficiently small that we are not forced to take it into consideration. This is a major difference between our everyday lives and our role as students of science. In science, the uncertainty is *always* significant, and research results are not considered complete without a full discussion of the uncertainties associated with them.

The reality of uncertainty.

The significance of uncertainty.

Significant Figures

In the previous chapter we learned to express the *magnitude* of a scientific measurement or calculation, in this chapter we will learn to consider the *reliability* of such a number. The magnitude of a scientific measurement is expressed in scientific notation. For instance, 4.67×10^{-3} is the numerical representation of a magnitude. In science, the reliability of a measure is typically expressed by the number of digits in the measurement's magnitude. The number of digits needed to accurately

Magnitude and reliability of measurements.

reflect a measure's reliability is known as the number of significant figures in the measurement. Correct representation of the number of significant figures in a measurement or calculation is often a stumbling block for beginning science students.

An example of significant figures in the measurement of time.

Typically, the number of significant digits in a scientific measurement is a function of the sensitivity of the measuring device. Consider the following example: Two people witness a minor traffic accident, the investigating police officer asks each at what time did the accident occur. The first witness has an analog watch with hands and responds "about a quarter to twelve." The second witness has a digital watch and responds "a few seconds after eleven forty-two." Clearly the two witnesses observed the accident at the same instant, and assuming they both checked their watches immediately after the accident, we can attribute the differences in their responses to the sensitivity of their particular measuring device. In numerical form, their estimates of the time of the accident could be represented by: 11:45 and 11:42.12.

In the case of the first witness, a scientifically-minded police officer considers the first three of the four digits as significant. That is, that witness is confident of the 11 and the 40 part of the estimate, but there is some uncertainty as to the time the accident occurred. It could have been 11:41 or it could have been 11:48, there is some degree of uncertainty as to the *exact* minute.

In the case of the second witness, the police officer can consider both the hour and the minute estimate to be certain. However, there is some uncertainty in the exact second at which the accident occurred, it could have perhaps been 11:42.05 or 11:42.29. The second witness was able only to record and report the time of the accident to the nearest minute. Thus, this second estimate would contain four significant figures, one greater than the estimate given by the first witness.

When reporting the results of a measurement or calculation, scientists can utilize the concepts of scientific notation and significant figures to represent both the magnitude and reliability of a number. Take for example the following list of values:

Mass of the Earth — 5.97×10^{24} kg
Density of Quartz — 3.65 g/cm^3
Average Distance to the Moon — 3.844×10^5 km

Clearly each measurement consists of three important pieces of information, the magnitude, the reliability, and the units of measure. From the expression of these measurements what can be said about their reliability? The mass of the Earth is expressed to three significant figures. That is, there is absolute certainty that the Earth's mass is at least 5.9×10^{24} kg, but we may expect as much as 1×10^{22} kg of error around the reported value 5.97×10^{24}. Thus, the possible range in mass is 5.96 to 5.98×10^{24} kg. Likewise, the density of quartz is known to three significant figures, and as such could range from 3.64 to 3.66 g/cm^3. The average distance to the Moon is known with a reliability of 4 significant figures, and thus may be in error by 0.001×10^5 or 1×10^2 km.

Ranges of uncertainty.

Commonly in science we combine several measurements in some form of mathematical equation to arrive at a desired expression. For instance, the density of a mineral such as quartz is expressed as the ratio of its mass to its volume (g/cm^3). Thus, density is a derived unit of measure, as discussed in Chapter 1. To assess the uncertainty associated with an estimate of density, it is necessary to consider the errors associated with the measurement of both the mineral's mass and volume. The strength of certainty in a calculation is akin to the strength of a length of chain. The entire calculation is no more certain than its most uncertain measurement — just as the chain is no stronger than its weakest link. There are two important rules for the identification of the proper number of significant figures to express in a calculation.

For multiplication or division, the answer is to be expressed with as many significant figures as the most uncertain measurement.

Rules for significant figures in mathematical operations.

For addition or subtraction, the answer is to be expressed to the same number of decimal places as the least certain measurement.

Multiplication and Division Examples

$$(5.421 \times 10^3)(4.5 \times 10^2) = 2.4 \times 10^6$$
$$(2.7795 \times 10^4) \div (2.816 \times 10^3) = 9.870$$

Addition and Subtraction Examples

$$3.542 + 1.13 = 4.67$$
$$1.198 \times 10^4 - 1.49 \times 10^2 = 1.183 \times 10^4$$

Often scientific calculations involve the use of conversion factors. These numbers are typically expressed as ratios such as:

Absolute numbers.

$$\frac{100 \text{ cm}}{1 \text{ m}} = 1$$

Importantly, with regard to the concept of significant figures, these numbers are considered to be constants with infinite precision. That is, there is exactly 100 centimeters in a meter, a number which is known to an infinite number of decimal places. Thus, with respect to the appropriate number of significant figures with which to report the results of a calculation involving a constant the conversion factor is not considered. For example consider the following simple problem: A river is found to be flowing at an average rate of 2.74652 km/hr, what is the rate of flow in centimeters per second?

Example of unit conversion using absolute numbers.

$$(2.74652 \text{ km/hr}) \times (1000 \text{ m/km}) \times (100 \text{ cm/m}) = 2.74652 \times 10^5 \text{ cm/hr}$$

$$(2.74652 \times 10^5 \text{ cm/hr}) \times (1/60 \text{ hr/min}) \times (1/60 \text{ min/sec}) = 7.62922 \times 10^1 \text{ cm/sec}$$

Thus, the answer is reported to six significant figures, because this is the level of uncertainty in the original measurement. The fact that the conversion factors are not carried out to 6 significant figures does not in any way affect the uncertainty of the calculation, which in this hypothetical example is extremely small for a measurement of this type!

Precision vs. Accuracy

The words accuracy and precision are commonly used in every-day speech as synonyms. However, in scientific usage they represent different but interrelated concepts. While perhaps subtle, the difference between accuracy and precision is of critical importance in the sciences.

Definition of accuracy.

Accuracy is the degree to which a measurement records the "true" value of the natural phenomenon under consideration. The concept of truth in science is a very slippery one, with which scientists are constantly forced to deal. For now, however, let us assume that any given natural phenomenon has associated with it one or more absolute magnitudes which could conceivably be known with infinite certainty. The uncertainty associated with inaccuracy is simply how far the measure is from that absolute value. Errors of accuracy are associated with human errors and mistakes, as well as malfunction of the measuring device.

Definition of precision.

A measurement's precision is a reflection of its reproducibility. One of the fundamental tenants of the philosophy of science is reproducibility. Verification of experimental results by reproduction lies at the foundation of modern science. A measurement that when repeated, under similar conditions, returns nearly the same value is said to be precise. An example which is commonly given to illustrate the differences between accuracy and precision is that of a dart player. If the object of the game is to score a bull's-eye with each toss, the accuracy of the player is the average distance between the darts and the bull's-eye. The precision of the player is the average distance between the three darts.

Precise but Inaccurate | *Accurate but Imprecise* | *Precise and Accurate*

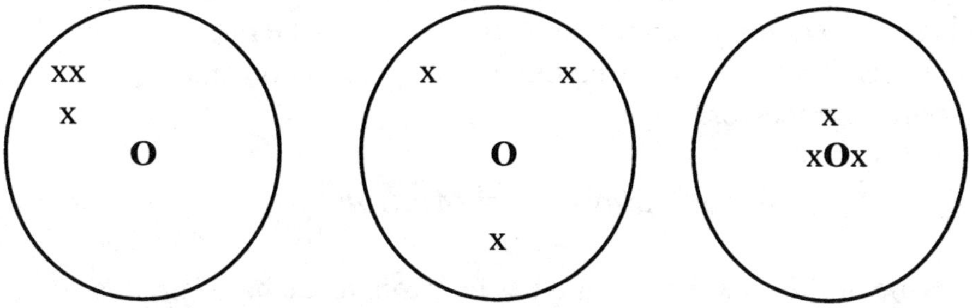

Examples of precision and accuracy in a game of darts.

Thus, it is very possible to be precise, while being inaccurate. Occasionally experimental data is found to be of otherwise high quality, but which is far from the "true" value. Such results are precisely wrong, and are often the source of stunning embarrassment to the scientists involved.

Because of the related problems of accuracy and precision it is of great importance that scientists — and the literate public in general — maintain a healthy skepticism towards all scientific results until clearly verified by reproduction. As scientists, the only thing we know with certainty is that we cannot know any scientific measurement with complete certainty. Scientists strive to reduce uncertainty by improving the accuracy and precision of all scientific measurements.

Random vs. Systematic Errors

Scientific measurements are never devoid of error. What are the common sources of such error? Typically, measurement errors can be ascribed to one of two general types, random or systematic.

Definition of systematic error.

Systematic errors are typically errors of accuracy. The measuring device is not calibrated correctly, or some similar problem leads to highly reproducible results which are found to be systematically in error from the known or expected value. Such errors are identified by making repeated measurements of well-characterized materials — analytical standards — under conditions identical to those encountered during measurement of unknown samples. Systematic errors can then be corrected for by applying a constant correction factor to bring the measured standard value into line with its known value, or if necessary by mechanical or electronic modifications to the measuring device.

Definition of random error.

Random errors are much harder to identify and correct. Typically, random errors are associated with errors of precision. One would identify the expected range of random errors by measuring a series of standards of constant composition, and evaluating the range of variation observed. Such errors are generally caused by human errors of measurement and observation or by temporary malfunctioning of the measuring device. While systematic errors can be eliminated by "tweaking" the machine or applying a correction factor, random errors can only be minimized by extreme attention to detail during the sample preparation and measurement processes.

Minimization of error.

Factor-Label Method

One of the most common types of errors made by beginning science students involves the setting up of a calculation and the elementary arithmetic necessary to solve it. Because these types of errors are not exclusively the property of students, professional scientists have developed a simple yet powerful technique to double-check their calculations and at least partially assure themselves of accurate results. This technique is know as the factor-label method, the essence of which is that units of measure can be treated as mathematical factors and combined by multiplication or canceled by division. The following example illustrates how the factor-label method is used to ensure the correct calculation of a conversion:

Calculate the velocity of a river in meters per second if the current is moving at 1 mile per hour.

Example of algebraic calculation using the factor-label method.

1 ~~mile~~/hr (x) 1.609 km/~~mile~~ = 1.609 km/hr

1.609 km/~~hr~~ (x) 1/3600 ~~hr~~/sec = 4.469 x 10^{-4} km/sec

4.469 x 10^{-4} ~~km~~/sec (x) 1000 m/~~km~~ = 4.469 x 10^{-1} m/sec

Do not forget the correct significant figures...

4 x 10^{-1} m/sec

Thus, by keeping close track of the units involved in the calculation, it is possible to be sure that no errors have been made. While no method is totally foolproof, this technique provides a powerful way to double-check for errors associated with the set up of a problem. It is always best, when dealing with a new type of calculation or conversion, to assure yourself that the mathematical steps you have set up result in the correct units prior to actually plugging in the numbers.

Chapter 2 - Exercises

Name ______________________ Date ______________________

Instructor ______________________ Section ______________________

1) Carry out the following calculations; report your answer with the proper number of significant figures.

a. $(2.45)(1.69 \times 10^{-3})$

b. $45 + 1.45$

c. $(8.4692 \times 10^{4}) \div (3.43 \times 10^{-2})$

d. $[(1.3356 \times 10^{2}) + (1.74 \times 10^{2})] \div (1.1156 \times 10^{4})$

e. $[(6.43) - (4.67 \times 10^{-1})][9.975 \times 10^{2}]$

f. $[4.47 \times 10^{2}][(3.3353 \times 10^{2}) - (1.17 \times 10^{1})]$

g. $1.756 \times 10^{6} + 12$

h. $(8.32 \times 10^{-4})(7.1 \times 10^{-3})$

i. $[(1.19 \times 10^{2}) + (1.16 \times 10^{1})][(4.63 \times 10^{1}) - (2.37 \times 10^{-1})]$

j. $(2.4416) \div (1.2208)$

2) a. Write a brief paragraph defining the concepts of accuracy, precision, and random and systematic error. Provide a mathematical basis for the measurement of accuracy and precision. Discuss why the method you describe is correct.

b. As part of an experimental procedure, the following analyses were made of a known laboratory standard. The standard had a "true" known and expected value of 1.4653×10^{-2}. Calculate the ranges of errors associated with accuracy and precision. Do these experimental results indicate random or systematic errors are a more important source of uncertainty in this case? Discuss your answer.

Trial	***Measurement***
1	1.4453×10^{-2}
2	1.1652×10^{-2}
3	1.6538×10^{-2}
4	1.4652×10^{-2}
5	1.5001×10^{-2}
6	1.4698×10^{-2}
7	1.3345×10^{-2}
8	1.4441×10^{-2}
9	1.3985×10^{-2}
10	1.4928×10^{-2}

3) Using the factor-label method, set up equations to solve the following problems. Note that you need not solve them; rather, set them up so that the proper conversions and calculations could be made.

a. The velocity of a river is measured in meters per second. What distance in miles has it traveled after 1 hour?

b. The mass of an object is measured in kilograms, its volume in liters. What is its density in grams per cubic centimeter?

c. Light from a distant star takes several years to reach the Earth. What velocity has the light traveled in centimeters per hour?

d. The rate of motion of the North American plate is measured in centimeters per year. How far will the plate have traveled in miles after $1x10^6$ years?

4) For each of the problems in Question 3, provide a number for the known quantities expressed to the proper number of significant figures. Solve each problem using the equations you worked out in Question 3 and the conversion factors in Appendix I. Express your answer to the proper number of significant figures.

a. __________ m/sec (3 significant figures)

__________ miles

b. __________ kg __________ lt (4 significant figures)

__________ g/cm^3

c. __________ years (5 significant figures)

__________ cm/hour

d. __________ cm/year (3 significant figures)

__________ miles

Chapter 3 - Variables: Dependence and Independence

Introduction

Science is not static. While a significant part of the scientific enterprise is directed at understanding what is, a more important pursuit is the discovery of how things change. The most common types of changes considered by the Earth sciences are changes in the nature or condition of something with respect to position or time. Spatial variation typically reflects change in horizontal or vertical position, while temporal change records evolution through time. In order to deal with such dynamic conditions scientists have developed the notion of functional relationships between variables. In this chapter we will explore the nature of these relationships, and come to understand some of the graphical conventions used to display the ever changing state of our world.

The Variable

The observational aspects of scientific study consist of the recognition and recording of natural phenomenon. Scientists however do not make a practice of recording their feelings or impressions with regard to these observations, rather the information that is recorded typically takes on some quantitative form. As we learned in the first two chapters, scientific measurements involve numerical values that have associated with them some level of uncertainty. To this point we have focused on developing an understanding of a scientific measurement as a numerical representation of a constant condition, yet scientific inquiry extends far beyond the analysis of constants. Natural phenomena are dynamic, that fact is pervasively observed in our daily lives; change in the weather can be sudden and violent, the tides rise and fall in subtly complex rhythms, the seasons change as the Earth revolves around the sun. In order to capture and quantify these changing conditions scientists have developed the concept of a variable.

Science evaluates static and dynamic phenomena.

Variables are, as their name indicates, variable. They record change, and they are changeable. The essence of the concept of a variable lies in the notion that an abstract symbol can represent all possible values of a measurable condition. By collapsing that infinite variability to a single symbol, it is possible to manipulate with great ease the changes it represents within mathematical expressions.

Definition of a variable.

For instance consider two small streams converging, the confluence of which produces a larger river. Each stream has some amount of water it carries that changes over the course of days, seasons, and years. No matter how much water flows through either channel, the two tributaries always combine to form the larger river. What then is the amount of water carried by that river? How can that value be expressed? It is possible that one could attempt to catalog the flow of each stream at all times and add the values together to calculate the flow within the river. However, this is a very cumbersome and tedious type of calculation to perform. Is it possible to represent the addition of stream flows in the form of a mathematical relationship that captures the essence of the situation, without becoming burdened with the details of the flow at every particular moment? By considering the flow in the streams to be represented by a single symbol, a variable, it is possible to write the simplified relationship:

Mathematical combination of static variables.

$$S_1 + S_2 = R$$

where S_1, S_2, and R are variables representing the flow rates of the two streams and the larger river at any given time respectively.

As written, the variables in the above expression are still static. They are said to represent the conditions of the rivers at any given time, but they still do not express how that flow rate varies from time to time. In order to capture that variability completely we need to understand how variables can be used to express change.

The Function

Following from the example above, does there exist a relationship between stream flow and time that is expressible in mathematical form? In this case there are two important variables, the flow rate of the given stream and time, represented by the symbols S and t respectively. The relationship between these two variables is one of a function. A function expresses the value of one variable as the other changes. In this case stream flow is a function of time. As time changes, stream flow changes. The symbols for this relationship are generally written in this form:

Definition of a function.

$$S = f(t), \text{ or } S(t)$$

Formulation of a functional relationship.

and would be read "stream flow is a function of time."

What is the meaning of this functional relationship? For each and every time t, there exists a unique and measurable stream flow S. Consider the decrease in stream flow after a flood. At its peak, the flow in the stream might have been 100 cubic meters of water per second (100 m^3/sec). Over an hour's time, the river was found to have decreased in flow rate by 10 m^3/sec. Since there are 3600 seconds in an hour, the functional relationship during the waning stages of the flood between time in seconds and stream flow in m^3/sec can be expressed by the relationship:

$$S(t) = 100 - (10 / 3600)t$$

Such a formulation is perhaps the simplest we could imagine for decreasing flow in a river after a flood. You likely recognized the form of the expression as that of a line with negative slope.

What is there to be learned about functional relationships beyond this elementary example? The most significant things to realize from the mathematical formulation presented above are that the two variables are linked by a relationship, and that such a relationship has associated with it some concept of cause and effect. As time passed, the flow in the river decreased. The way in which the formula is written describes the cause and effect of that relationship. Stream flow is a function of time, time is not a function of stream flow.

Functions as indicators of cause and effect.

Independent and Dependent Variables

Variables are used to represent any changing condition, however they are typically grouped into one of two types, the independent and the dependent variables. Independent variables are those which are considered to vary in a systematic way that is not influenced by changes in other variables. Because of this independence, they serve as the basis for the analysis of variation of other variables. Conversely, dependent variables are those wherein their value depends directly upon the value of another variable. This dependence is the cause and effect relationship discussed in the case of the waning flood.

Definition of independent variables.

Definition of dependent variables.

Typically, independent variables are ones of time or position. The other variables, those that depend upon the value of some other variable, are measured at different times or places and are then represented by a mathematical expression that describes how they change as a function of their dependence. From the stream example, it is obvious that the river's level changes as time passes. The passage of time can be measured independently of any other condition. The seconds flow from one to the other in an uninterrupted sequence, the passage of which marks, but is not altered by, other variables. There is, however, no way to express the changes in the level of the stream without referring to the passage of that time. It is perhaps too strong to state that the passing of time causes the river to recede, but it is mathematically convenient to express it that way:

$$S = f(t)$$

The functional relationship is said to map the changes in the independent variable onto those of the dependent variable. Each incremental change in the independent variable results in a calculable change in the dependent variable. This concept of a functional relationship between variables is extremely powerful, it allows us to grasp the cause and effect of natural phenomenon, as well as describe mathematically those observed changes in terms of a framework that is external to the dynamic system in question.

The importance of cause and effect.

The recognition of functional relationships, the understanding of cause and effect, is central to the study of dynamic systems. Nature's complexity makes clear recognition of fundamental functional relationships underlying complex systems very difficult at times. As a student of science you should always endeavor to grasp the nature of functional relationships as they are discussed. Ask yourself which of the variables being considered is independent and which is dependent? What is the form of the functional relationship? Is it increasing, decreasing, or changing in an oscillatory fashion? Perhaps most importantly, you should always be searching for the underlying cause and effect relationships that give rise to the observed variation? Only by approaching natural systems in this way will you begin to understand and appreciate the complex causal relationships that exist between dynamic systems.

Graphical Conventions

The simple elegance and wide applicability of the concept of a functional relationship allows for its expression in several different ways. For compactness no representation is better than the mathematical symbols held within a formula. However, it is sometimes difficult to grasp the nature of the relationship under consideration solely from the functional form. Consider the cooling of a flow of lava which has erupted from a volcano. As heat is radiated from the lava into the atmosphere and underlying rocks the temperature of the flow might be found to decrease according to the function:

$$T(t) = T_0(t + 1)^{-1}$$

Graphing functions.

where T_0 is the initial temperature at time 0 and t is time in days. What is it that this formulation represents? For one who is inexperienced in mathematical relationships of this type, it might be difficult to grasp the underlying structure of the time-temperature cooling path. However that functional relationship may also be represented in the form of a graph:

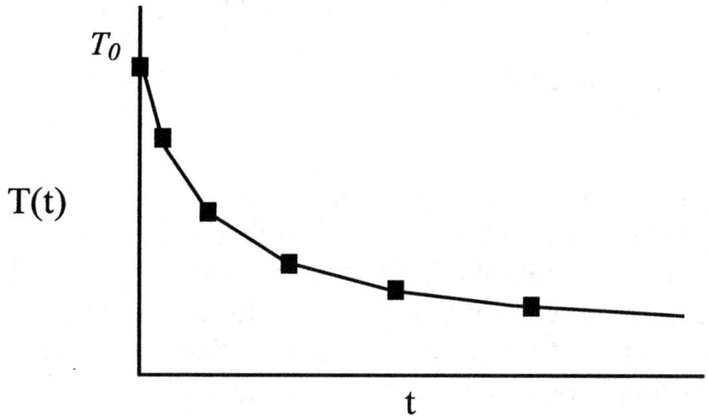

Graph of a decreasing function.

What can we learn from this graph that may not be immediately obvious from the formulation? First, the temperature of the lava decreases from the initial constant T_0 with increasing time. Second, the rate of that decrease is not constant; the graph is one of a curved line, not a straight line. Third, the rate of decrease in temperature decreases with increasing time. That is, the curve flattens from left to right. Clearly, graphs of functions provide an abundance of information that may not be immediately obvious when the functional relationship is only considered in its symbolic form.

Observation of a graphical representation of a function.

Importantly, the graph of the time-temperature function is constructed in a particular way. The horizontal axis (abscissa) is used to represent the independent variable, while the vertical axis (ordinate) is used to represent the dependent variable. The reason behind such a graphical convention is perhaps made most clear when we consider the construction of the graph from a table of data calculated from the formula:

Rules for placing independent and dependent variables on graphical axes.

$$T(t) = T_0(t + 1)^{-1}$$

time (days)	*temperature (units of T_0)*
0	T_0
1	$T_0/2$
2	$T_0/3$
3	$T_0/4$
4	$T_0/5$
5	$T_0/6$

To construct such a table one would choose a value for t (one, two, or three days) and then utilize that value to calculate the temperature at that given time. When graphing the results of these calculations, it is most convenient to divide the horizontal axis into regular increments of the independent variable (typically equivalent to the values at which the function was evaluated). The calculated value of the dependent variable can then be plotted at the appropriate location on the graph with ease.

Establishing causal relationships from a graph.

Almost without exception in the sciences, functional relationships are graphed with the independent variable along the horizontal axis and the dependent variable along the vertical. When observing the graph of a function in the absence of the equation in functional notation, one can assess the nature of the cause and effect relationship between the two variables simply by their positions on the graph.

Chapter 3 - Exercises

Name ________________________ Date ________________________

Instructor ____________________ Section ______________________

1) Explore one of the World Wide Web sites provided by your instructor. Record at least three functional relationships described in that site. Define the form of the function and variables used in each relationship.

Site Name : ______________________________

Relationship	Function	Variables
1		
2		
3		

2) Describe five everyday occurrences that are functions of time. List the variables and define the nature of the functional relationship. Refer to the list of suggested questions to ask yourself from Chapter 3.

I.

II.

III.

IV.

V.

3) The velocities of rivers are found to decrease with increasing distance of flow. That is, high mountain streams are generally fast flowing, while rivers near the ocean are much more sluggish. Define the functional relationship between velocity and position. Which is the dependent and which is the independent variable?

If the function is found to be curved, with rate of river flow decreasing with increasing distance, sketch the form of the graph in the space below. Define the axes according to convention.

4) Consider the passage of a summer rainstorm. The intensity of rainfall changes from no rain to a drizzle to a hard rain and then back again until the storm is passed. Define the important functional variables in this relationship.

In the space below sketch the form of the rainfall intensity over the course of the passage of the storm; label the axes appropriately.

5) The force of gravity between two objects is found to vary directly as a function of their masses, and according to an inverse square rule as a function of their distance of separation. Write an equation for this relationship that uses the masses of the objects and the distance between them to define the gravitational attraction. Which are the dependent and which are the independent variables?

Imagine that the masses of the two objects are identical and constant. Sketch the graph of the functional relationship between gravity and distance in the space below.

Chapter 4 - Graphs: Form and Function

Introduction

Quantitative data serve as raw material for the engine of scientific progress. The collection and analysis of data form the foundation of experimental and empirical scientific study. Likewise, consideration of relationships within and among data sets defines the realm of the theoretician. As such, irrespective of the specific nature of a scientist's work, acquisition, examination, and presentation of data are necessary for the study of the natural world. Previous chapters have considered the basics of numerical measurements and functional relationships. This chapter provides an introduction to graphs and the graphical analysis of data. While perhaps daunting at first, graphs are nothing more than pictures used to illustrate concepts and relationships that might otherwise take pages of written description to make clear. The ability both to construct and comprehend graphical data is central to your understanding of science and will prove to be a skill that will serve you well , irrespective of your chosen career.

Continuous and Discrete Data

The evaluation of numerical data generally occurs within a context of space and time. In both cases, populations of data are found to define either of two general forms. Data that describe smoothly varying and unbroken functions are said to be continuous, while data that are incremental are said to be discontinuous. For example, the length of wire along a fence line is continuous and unbroken as it stretches from post to post. Any point along the wire can be defined within the limits of the precision of the measurement. Conversely, the positions of the fence posts are discrete; each occupies a specific location that is separated from its neighbors by a fixed distance.

Definitions of continuous and discontinuous.

Mathematics has established a formal definition of the concept of continuity, a thorough treatment of which is beyond the scope of this book. However, for our purposes, it will be necessary to construct a working definition that we can apply to natural data. For many systems, the boundary between discrete and continuous is blurry, and highly dependent upon the scale of observation. Consider the dots of ink that make up a newspaper photograph, or the pixels of color that define images on computer monitors and television tubes. When viewed at a coarse scale, as from a sufficient distance, the discrete cells of color blend together to define a continuous image. However, when viewed at high resolution, such as under a magnifying glass, those continuous images are found to actually be composed of individual spots. Unlike mathematics, science affords us the opportunity to consider data as either continuous or discrete depending upon the application in question.

Continuity can be scale dependent.

Continuous mathematical function.

When we think of mathematical relationships, we typically have in mind continuous functions. For instance the equation:

$$f(x) = 2x + 4$$

is continuous over all values of x.

Domain and range of a function.

From mathematics, we learn that the set of values that the variable x can assume is known as the domain of the function, while the set of values that the function f(x) takes on is known as the function's range. Many natural systems are found to be continuous across their domain. For instance, the surface of a quiet pond is continuous. The flat surface defines a smooth and unbroken plane.

Time is a continuous variable.

Perhaps the most fundamental continuous function is time. While humans have for convenience subdivided time into a series of discrete hierarchically arranged subdivisions such as years, months, weeks, and days, the underlying continuity of time is unquestioned. Because of this continuity and functional independence, time is most commonly chosen as the variable against which other changes are considered.

Graphs of continuous functions are found to define smoothly varying curves. For example, the function

$$f(x) = \sqrt{x}, \quad \text{for} \quad x >= 0$$

has the general form:

Graph of a continuous function.

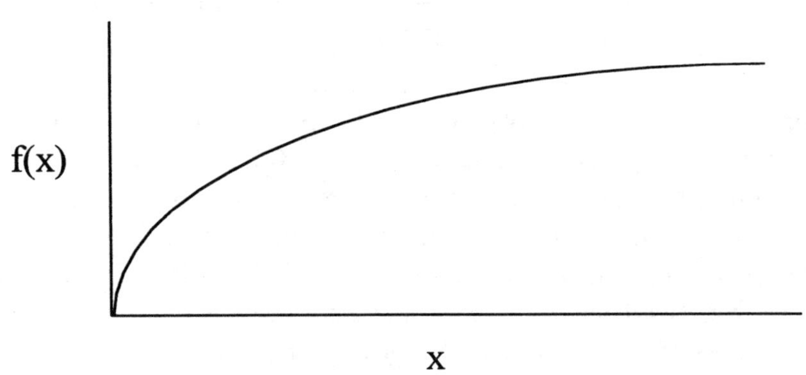

over the domain of positive real numbers and zero. Any value of x from that domain may be mapped into the function f(x). Thus, the function is said to be continuous over the domain x >= 0.

In nature, many systems that appear to be continuous at one scale are found to be actually discrete when viewed at a higher resolution. Consider the example of a beach gleaming in the tropical sunshine. When viewed from a distance, the sand appears to define a rippled surface that smoothly stretches continuously along the shore from one horizon to the other. However, closer examination reveals the beach's true character. The shining surface is actually composed of innumerable individual grains of sand. The reality of this discontinuity is experienced on a windy day, when the sting of sand grains is commonly felt against the beach goer's ankles.

Graphs of discontinuous data sets consist of individual points or perhaps short segments of curves that are separated from each other by discrete breaks or discontinuities. For example, consider the six-sided die commonly encountered in games of chance. If one were to roll the die 60 times, recording the outcome of each trial as the integer value represented by the number of spots on the upper surface, one might develop a graph such as this:

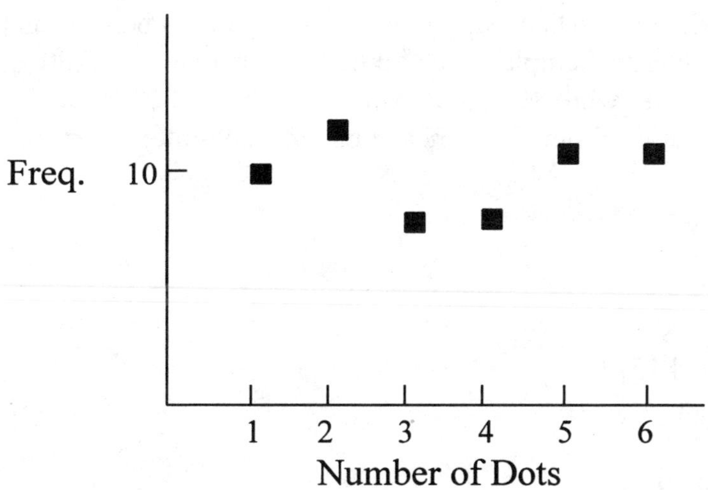

Graph of discontinuous data.

In this case, the data define a discontinuous population centered on the frequency 10. The independent variable exists as a set of discretely varying values (i.e. 1, 2, 3, 4, 5, 6). Likewise, the frequency of occurrence of any particular face may only be incremented at integer values. Clearly it is absurd to consider the notion of anything other than an integer number of trials or throws of the die. That is, there is no way that an experiment could result in $11^{1/2}$ 3's, or $8^{3/4}$ 5's.

Approximation

Collecting discrete data.

The collection of scientific data almost always consists in the tabulation of discrete measurements: atmospheric pressure is measured at specific elevations across a mountain range, the lengths and widths of individual cobbles are tabulated, the discharge of a river calculated at daily increments. From these discontinuous data scientists are able to make inferences as to the underlying functional relationships between variables. That is, scientists compile discrete data in order to approximate a presumed continuous function.

Approximations are the ways in which scientists bridge the gaps between their discontinuously collected data. One of the most challenging intellectual tasks of a research scientist is the search for underlying patterns within discontinuous data. There are two forms of approximation commonly used to define a continuous function from discrete data: interpolation and extrapolation.

Definition of interpolation.

Interpolation is the process by which one fits a smooth curve or straight line segment between individual data points. For example, consider the relationship between mass and volume for a pair of samples of the mineral quartz. Sample 1 was found to have a mass of 5.00 g and a volume of 1.90 cm^3, while sample 2 exhibited a mass of 11.0 g and a volume of 4.15 cm^3. When these measurements are plotted the following graph results.

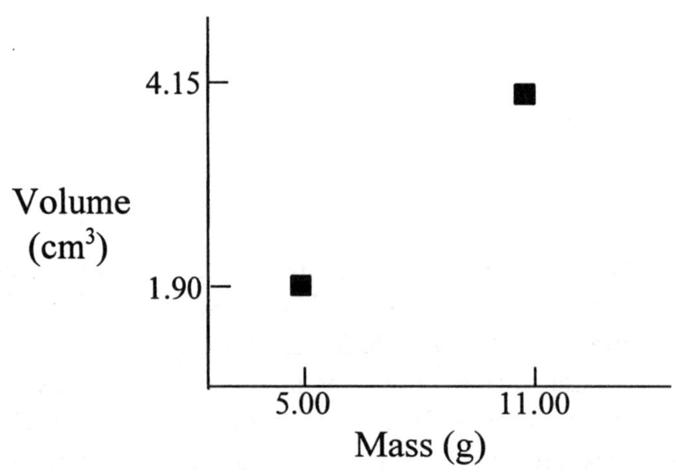

A line drawn between these two points suggests a constant relationship between mass and volume for the mineral quartz. We know this relationship to be a reflection of the concept of density. Thus, by interpolating a linear relationship between the two known points, it is possible to calculate the expected mass of a sample given a known volume (provided, of course, the unknown sample is also a crystal of quartz).

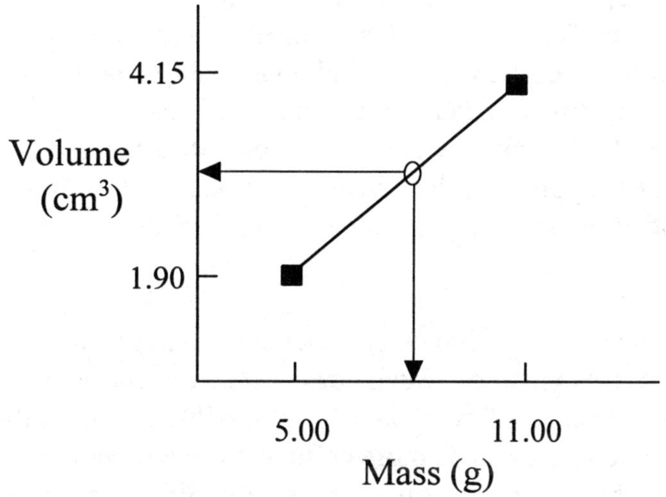

Graph of an interpolation.

The second type of approximation that extends continuity from discontinuous data is extrapolation. Extrapolations expand a functional relationship to portions of its domain for which there are no data. From the example of the quartz crystals discussed above it is possible to extend the density relationship beyond the observed measurements. Given a quartz crystal with volume of 15 cm^3, one could use the established relationship between the two measured samples to extrapolate the expected mass of the new sample.

Definition of extrapolation.

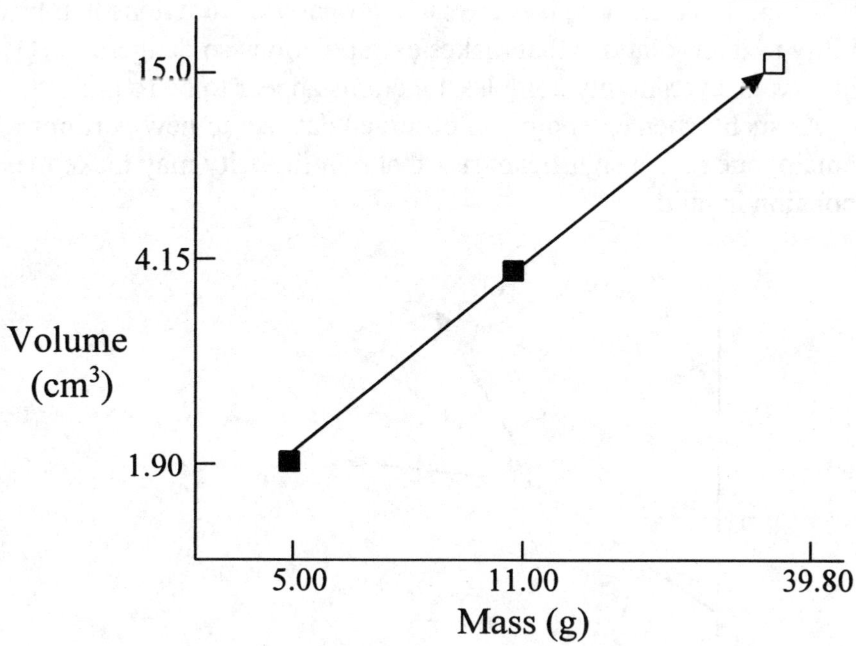

Graph of an extrapolation.

As illustrated by the graph, one would find the extrapolated mass to be 39.8 g. The question comes immediately to mind: How valid is this approximation?

The practice of approximation through interpolation or extrapolation is inherently dangerous from a scientific standpoint. It implies knowledge of the underlying functional relationship that is not directly borne out by the observed data. Where and when is approximation appropriate? Typically one has more confidence in interpolations than in extrapolations, for obvious reasons. An interpolation is bounded by observed data on either end of the approximation.

Sources of error in interpolation.

Errors in interpolation develop when functions are not well approximated by linear relationships. Such errors may be reduced and almost eliminated by collection of more data. It is possible to "fill in the gaps" within a data set to approximate more completely the function in question. Indeed, it is this notion of more finely subdividing a continuous function that led directly to the development of the differential calculus. However, one should remember that discontinuous data are always discontinuous, and while the collection of more data will decrease the distances between data points, there will always remain gaps over which approximations must be made.

Sources of error in extrapolation.

While interpolations may be made more robust by reducing the distance over which the approximation is extended, extrapolations are considerably more difficult to handle. An extrapolation is bounded only on one end, and as such, is less well-constrained than the typical interpolation. Interestingly, it is the same property of functions that lends credibility to interpolations that makes extrapolations so dangerous. Over short intervals, even highly complex functions appear to be roughly linear. As such, when extending an observed data set to new portions of the domain, one runs a significant risk that non-linearity may make the extrapolation invalid.

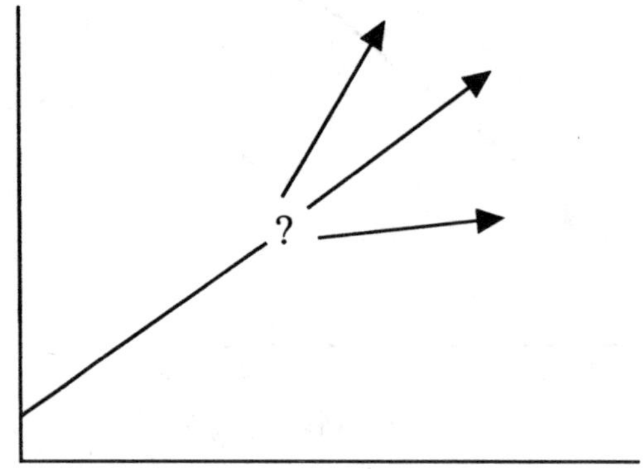

Clearly, the practice of approximation may only be applied with validity in those cases where sufficient data have been collected such that the approximation of the function may be done with reasonable certainty. The establishment of reasonable certainty lies at the heart of experimental and empirical science. Only through the careful collection of data, followed by thoughtful consideration, should a worker feel comfortable constructing meaningful approximations. It is the role of the theoretician to search out and account for the underlying functional relationships that are exhibited by discrete data, and as such, pave the way for legitimate interpolations and extrapolations.

Reasonable certainty in approximation.

Typical Graphical Forms

The form of a graph is directly dependent upon its intended use. It is critical to remember that graphs are constructed for convenience. They illustrate relationships that might otherwise prove difficult or impossible to observe within a set of data. Typically within introductory science classes two simple graphical forms are encountered: the scatter plot, and the histogram. Scatter plots are graphs that plot the value of one variable against that of another. The mass and volume of a crystal of quartz provide a so-called coordinate pair of data that are plotted on a set of perpendicular axes. As discussed previously, the variable that is considered independent is typically plotted on the horizontal axis (abscissa) while the dependent variable is plotted on the vertical axis (ordinate). For instance, the amount of water flowing down a river at any given time is commonly plotted on the vertical axis, with time as the horizontal axis.

Scatter plots of data.

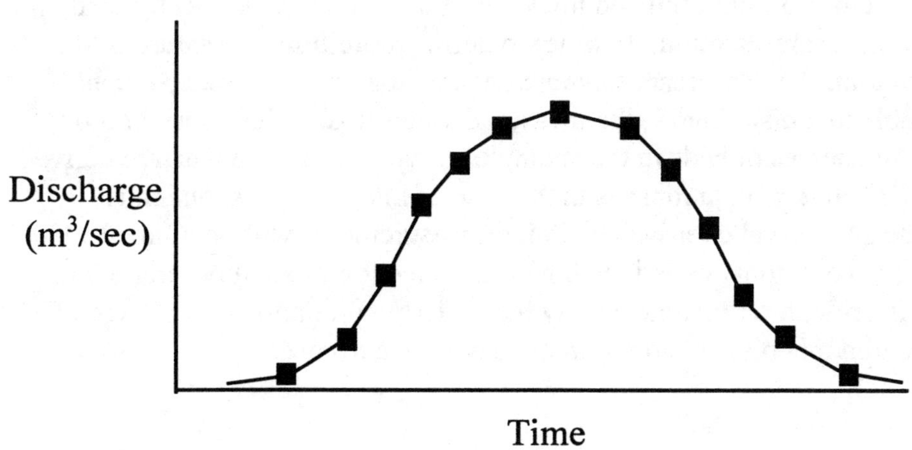

Example of a scatter plot graph illustrating discrete measurements of a continuous function.

Such a graph illustrates the change in discharge over the course of several days, and clearly indicates a period of flooding followed by a gradual return to more normal conditions. When investigating a new phenomenon, scientists typically plot their data in similar scatter plots in order to search out the presence of patterns or underlying relationships. A second common type of graphical representation of data is the histogram. These graphs illustrate the frequency of occurrence of data of various magnitudes. Such graphs are well suited to illustrate relationships of variation within large populations of data. Consider the observed thicknesses of individual layers (beds) of rock found in a vertical sequence. How could one represent that population of thicknesses? For a start, one would construct a scatter plot of bed thickness (dependent variable) versus bed number in the vertical sequence (independent variable).

Scatter plot of rock thickness.

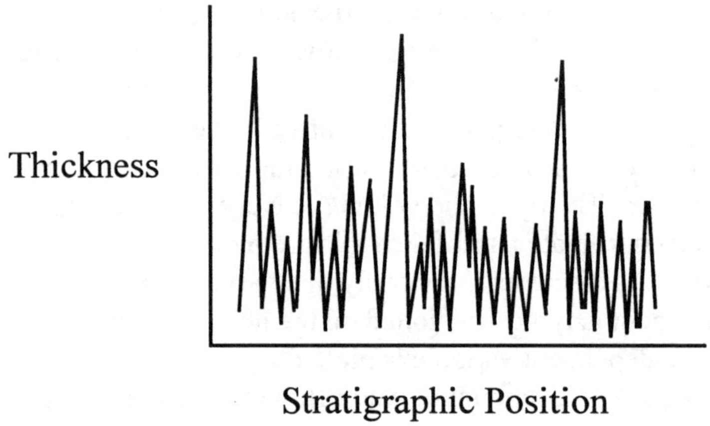

However, it is difficult to learn much about the nature of bed thicknesses in the population from this graph. One can identify the maximum and minimum bed thickness, but little more is directly evident from this representation. In order to define more fully the nature of the population, it is necessary to construct a histogram. The independent variable in a histogram is the measured magnitude under consideration, the thicknesses of beds in the example above. Unless the data are derived from discrete populations, as in the case of integer values, one must define an interval over which similar measurements will be lumped. This range is commonly called the bin size. Selection of an appropriate bin size is critical to illustrate clearly the underlying relationship between the measured magnitude and the frequency of occurrence.

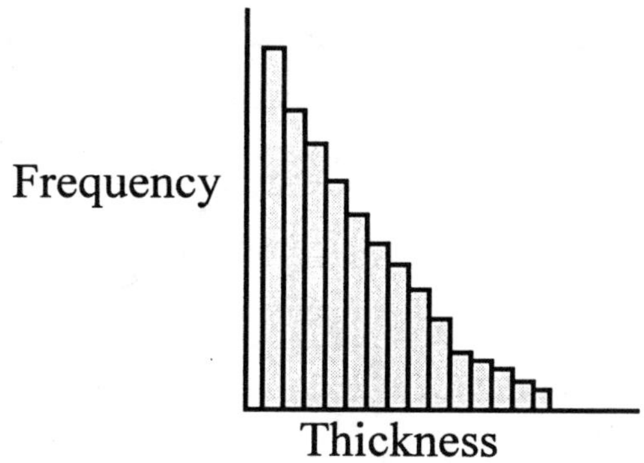

Histogram of discrete data.

When too coarse a bin size is selected, the entire population falls within only a few intervals, and as such fails to expose any underlying functional relationship between magnitude and frequency. Likewise, when too small a bin size is chosen, the population becomes too spread out and again the underlying relationship can be lost. However, when an appropriate bin size is chosen, the relationship between bed thickness and frequency of occurrence is well illustrated. One must always approach binned data with some degree of skepticism since choice of bin size can either mask or exaggerate a relationship within a population of data.

Role of bin size in the shape of histograms.

Correlation vs. Causation

Groups of variables that describe a relationship between individual pairs of values are said to be correlated. A formal definition of correlation will await the next chapter. For now, however, it is sufficient to say that correlated variables are those which vary in a systematic fashion. For instance, the volume of a mineral sample is correlated with its mass. As one increases the other does also in a systematic way. Correlations are either direct or inverse. Directly, or positively, correlated variables are those wherein an increase in one value results in a change in the other of similar sign. That is, increases result in increases and decreases result in decreases, such changes are represented by the mathematical relationship:

Correlation of variables.

$$A \propto kB$$

Positive correlation.

where k is the constant of proportionality. On the other hand, inversely, or negatively, correlated variables change in an opposite fashion. An increase in one variable results in a decrease in the other. This inverse relationship is written as:

$$A \propto k/B$$

Negative correlation.

Graphs of positively and negatively correlated functions.

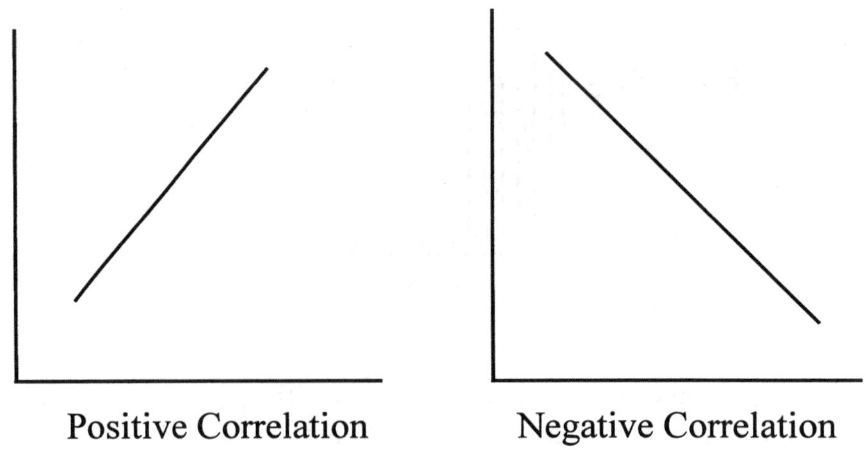

Establishing cause from correlation.

The presence of a relationship between two variables, either positive or negative, is generally taken as evidence of some form of causal relationship between the two. In the case of the mineral samples of quartz, we know mass and volume display a correlation, and that observed correlation defines a real underlying functional relationship. The constant of proportionality between mass and volume is density. That is, mass equals density times volume. The density of the mineral samples is constant, and mass and volume vary proportionally from sample to sample. It is not uncommon however in the sciences to find a correlated set of variables that in fact do not share a causal relationship. Throughout the history of science there have been many examples of errors of interpretation derived from the mistake of assuming a causal relationship between correlated variables when in fact no such causality existed.

Evaluating cause from graphical data.

It is critical for students of science, at all levels, to consider always the underlying assumptions of causality between variables. The presence of some form of correlation implies causality, but does not require it. As you evaluate graphed data, consider the implied or explicitly stated functional relationships between the two variables. Is the supposed causal relationship valid? Is it presented in a meaningful way? Are there underlying assumptions that are not fully expressed? Posing and answering these questions will improve your understanding of both graphical representations of data and the underlying scientific principles they are designed to illustrate.

Chapter 4 - Exercises

Name ______________________ Date ______________________

Instructor ______________________ Section ______________________

1) Provide five examples of natural systems you have learned about in your Earth science class which exhibit properties of continuity.

a.

b.

c.

d.

e.

2) Provide five examples of natural systems which exhibit properties of discontinuity.

a.

b.

c.

d.

e.

3) Provide three examples of natural systems whose continuity changes as a function of level of observation, much like the beach and its constituent sand grains.

a.

b.

c.

4) Sketch a graph of a continuous function.

5) Sketch a graph of a discontinuous function.

6) Using the following sets of coordinate pairs, calculate linearly interpolated values for f(x) given a value of x.

	x_1	$f(x)_1$	x_2	$f(x)_2$	x_3	$f(x)_3$
1	1	1	5	5	3	
2	-2	4	2	12	1	
3	1	4	4	-2	2	
4	-2	-2	2	-8	0	
5	4	4	7	4	6	
6	-4	-2	8	4	2	

Work Space:

7) Using the above coordinate pairs, calculate linearly extrapolated values for f(x) given the following values of x from either side of the pair.

	x_4	$f(x)_4$	x_5	$f(x)_5$
1	-2		8	
2	-4		6	
3	-2		8	
4	-6		6	
5	-10		27	
6	-16		24	

Work Space:

8) On the graph below, plot and label the following coordinate pairs.

x	-8	-7	-7	-5	-5	-4	-3	-2	-1	-1	0	1	1	3	3	4	5	6	8
f(x)	-2	6	1	8	-4	7	0	-2	5	6	1	4	3	6	-4	-1	-5	4	0

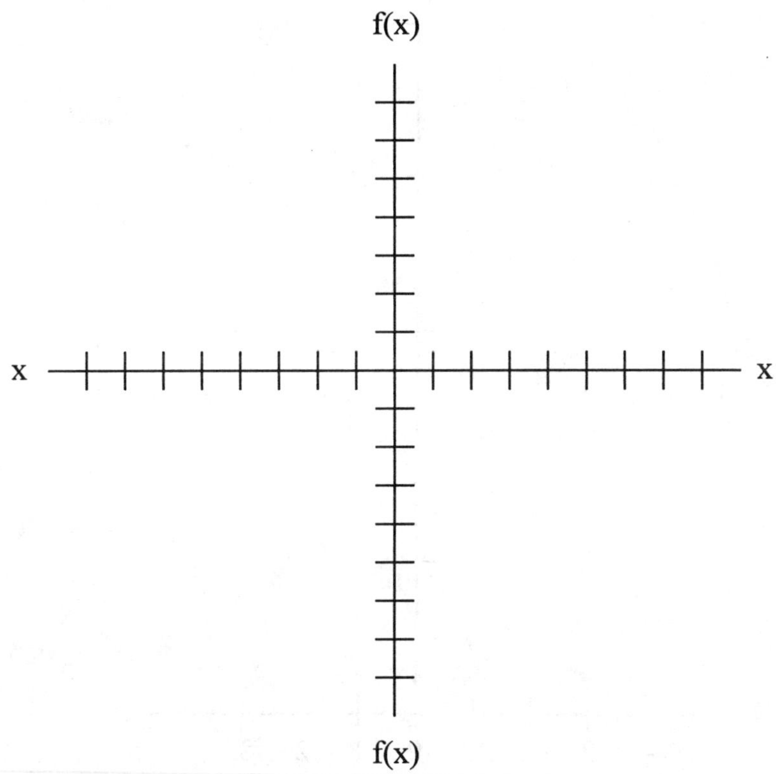

9) From the following catalog of values, calculate the frequency of occurrence given the bin sizes 1, 2, and 4.

-4, -4, -3, -3, -2, -2, -2, -1, -1, -1, -1, 0, 0, 0, 0, 0, 0, 1, 1, 1, 1, 2, 2, 2, 2, 3, 3, 3, 4, 4

Bin	-4	-3	-2	-1	0	1	2	3	4	5
Freq										

Bin	(-4, -3)	(-2, -1)	(0, 1)	(2, 3)	(4, 5)
Freq					

Bin	(-4, 0)	(1, 5)
Freq		

10) Plot the calculated frequencies from above as the dependent variable on the graph below. For bins covering more than one integer value, plot the point at the midpoint of the bin range.

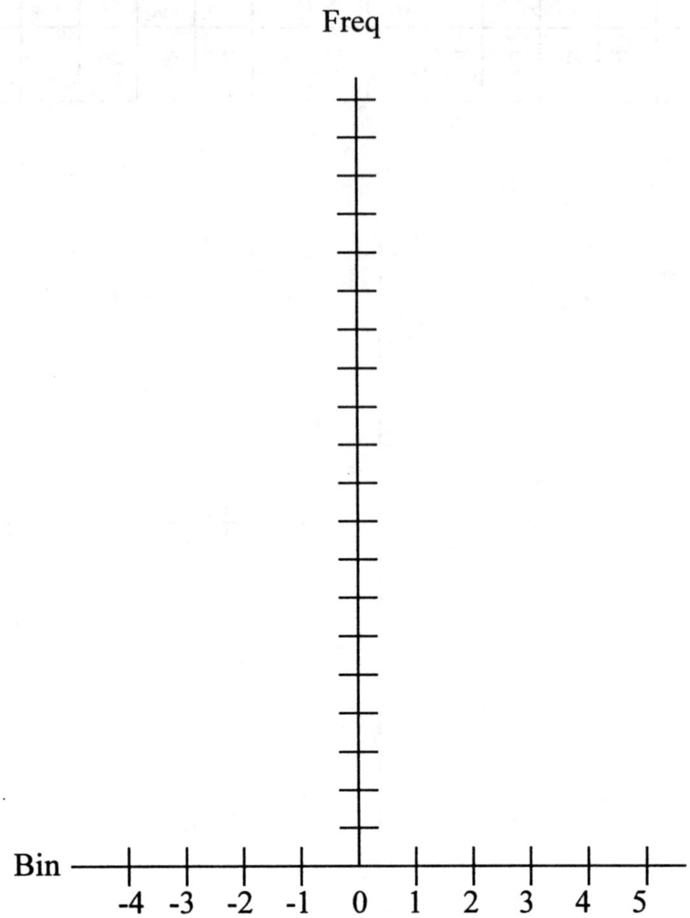

Chapter 5 - Graphs: Design and Construction

Introduction

The study of science requires more than simply the observation and interpretation of graphical data. A student of science must be able to create a graph that is meaningful and understandable. In order to correctly design and construct a graph it is critical that you understand the various graphical components, their purpose, and how to incorporate them into your graph. Throughout this chapter, a simple scatter plot graph is used as an example. However, it is important for you to realize that the principles discussed herein are applicable with only minor modification to nearly all forms of graphical data.

Axes and Their Quadrants

The basis of all graphs are the axes that define the graphical field. These axes serve as scales for the independent and dependent variables. Graphical axes are typically constructed perpendicular to each other, with one occupying a horizontal orientation and the other a vertical orientation. In almost all cases, the independent variable is plotted along the horizontal axis, and the dependent variable or variables are plotted along the vertical axis.

Axes are simply number lines along which the variables are positioned. The only significant difference between a graph's axis and a standard number line is that the value represented by a point on the axis extends perpendicularly from it. Consider the example below:

Graphical axes as number lines.

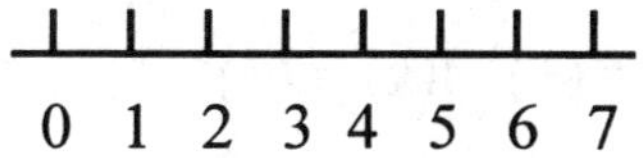

Number Line

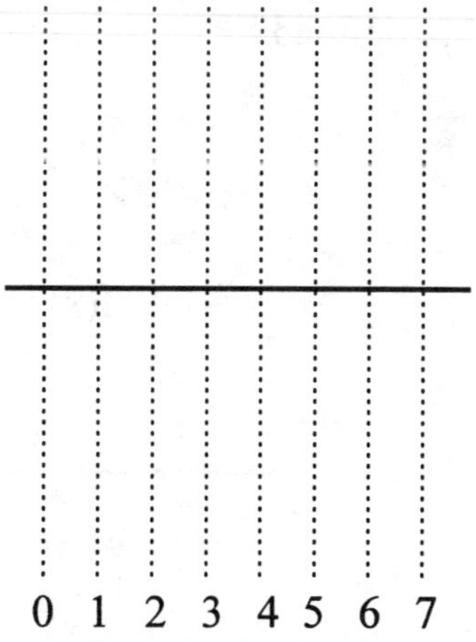

Horizontal Graph Axis

Because of the interesting ability of graphical axes to record the value of a variable above and below the actual line of the axis, it is possible to construct a set of perpendicular axes, and thus locate positions with unique values from two different variables. Typically, one constructs a horizontal axis extending from negative values to positive values. Then, a second vertical axis for the independent variable is constructed perpendicular to the first axis with the positions of the two variables' zero values coinciding at the point of intersection.

Independent and dependent variable axes.

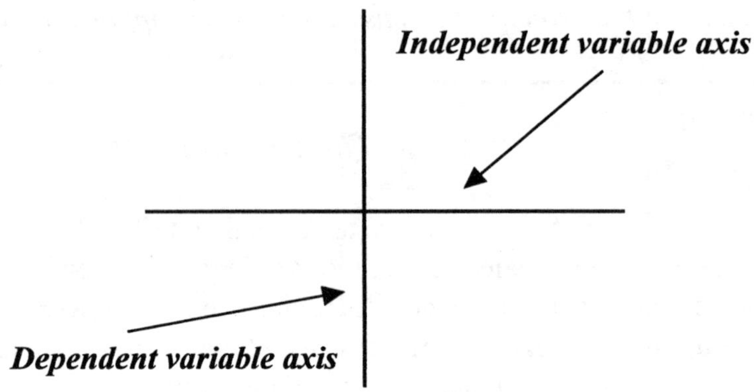

Once the field of the graph is defined by the two axes, it becomes clear that there are four quadrants of the graph. If the graph is constructed so that the independent variable x is plotted along the horizontal axis, and the dependent variable f(x) is plotted along the vertical axis, then the four quadrants are defined by the positive or negative value of both variables.

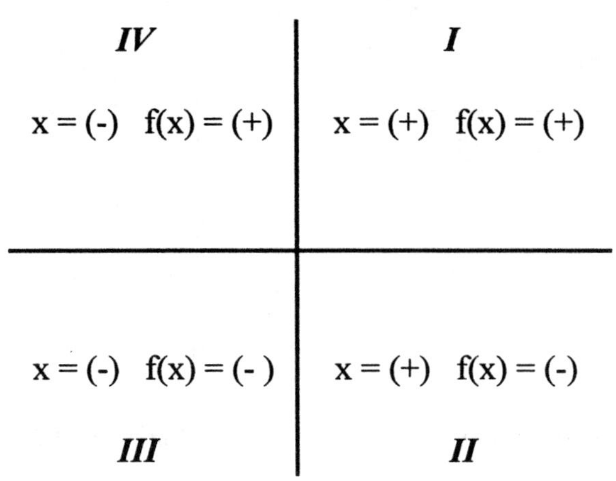

As illustrated, the upper right hand quadrant contains values of both x and f(x) that are positive; this portion of the graph is commonly called the first quadrant. The second quadrant, lower right field, is the region of positive values of x and negative values of f(x). The third quadrant (lower left) is the region of both negative values of x and f(x), while the fourth quadrant (upper left) represents negative values of x and positive values of f(x). One of the most common graphical errors encountered by students is simply plotting a coordinate pair in the wrong quadrant. It is critical to pay close attention to the sign of value of each variable, and double check that you have placed the point in the proper quadrant.

Scales

As pointed out above, the axis of a graph is simply a number line. Number lines are commonly constructed with every integer value marked and labeled.

Scaling on a number line.

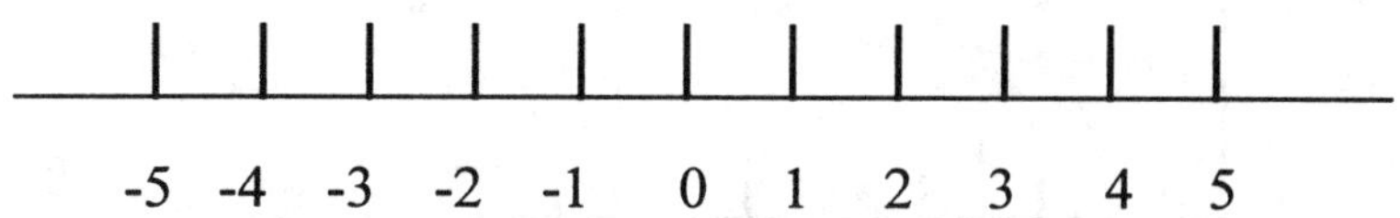

However, for convenience, it is sometimes useful to omit some of the integer values on a graph. For instance, if one were asked to graph a set of data the values of which ranged from near zero to over 100, it would cause the graph to be too cluttered to mark and label each integer value of the independent and dependent variables. Rather, it might prove more useful to mark positions along the axes at larger intervals; perhaps every fifth or tenth value could be labeled.

Labeling intervals along a graph's axis.

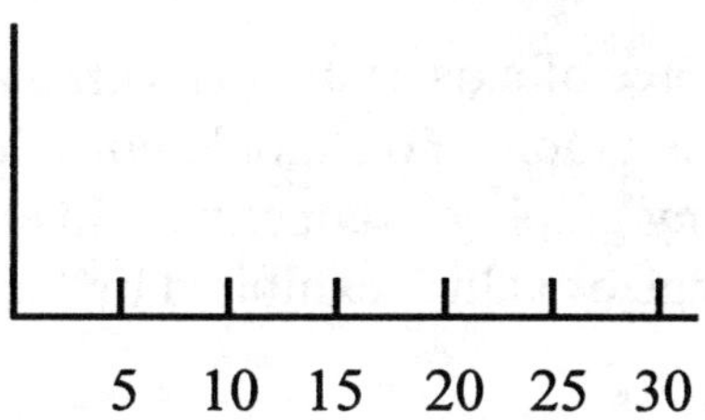

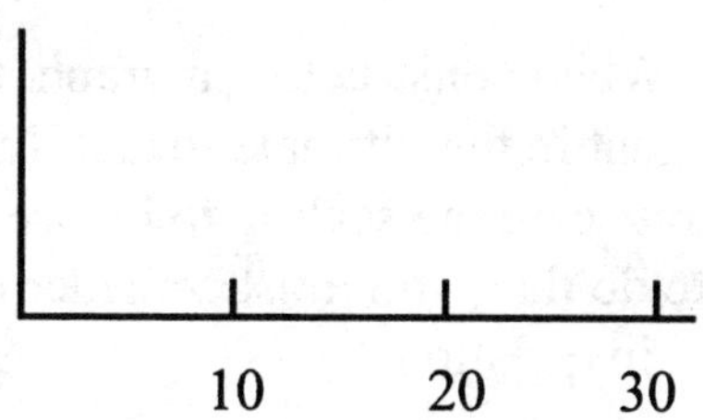

Conversely, it is at times useful to construct a graph that is a mix between the standard number line style and the incremented style. In these cases, the graph's axes are marked at integer values, with perhaps ever fifth or tenth value labeled.

Combination style number lines as graph axes.

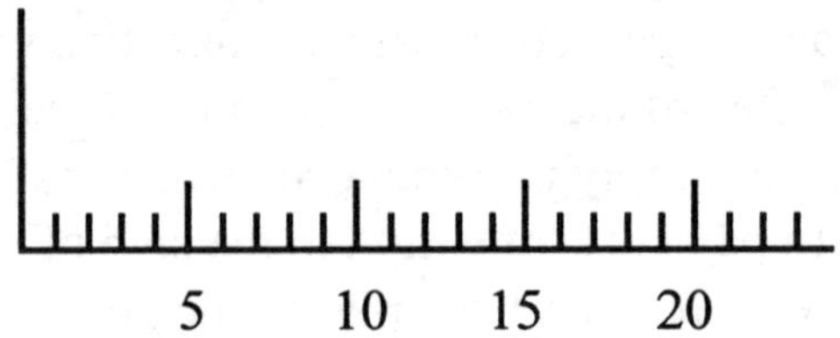

Finally, perhaps the most commonly encountered form of axes scaling is that wherein increments are marked and larger increments are labeled. For instance, perhaps every fifth value is marked and every twentieth value is labeled.

Doubly incremented graphical axes.

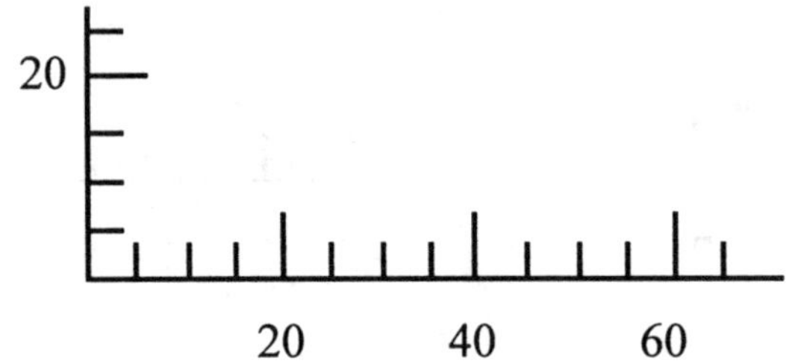

Linear vs. non-linear scaling of graphical axes.

Note that in all cases, no matter which form of scaling is used along the graph's axes, the distance between marked values has been constant. Graphs of this type are said to be linear graphs, because the axes scale in a linear fashion. It is not uncommon in the sciences to encounter more complex graphical forms that have axes scaled at non-constant intervals.

When constructing a graph, the choice of axis scaling is a critical component in the ultimate readability of the graph. You should consider with great care the scaling to be used on any graph you construct. In order to do that, you must consider the range of values exhibited by the data set in question.

Data Ranges

Sets of data have maximum and minimum values that define the range of the data. The range exhibited by a set of data directly controls your choice of scale when constructing a graph. Obviously, one must select a scale that allows all the data to be plotted. However, for utility and readability, one wants to construct a graph with a scale that allows all the data to be represented, while not leaving large regions of the graph devoid of data. Such large regions of empty space on a graph distract the viewer, and commonly reflect a lack of forethought or planning in the graph's construction.

Data ranges control graph scale.

If possible, one should attempt to choose a scale wherein the data largely cover the graphed field. Errors in choosing a good scale result in graphs that fail to display the data in the most useful way. Consider the following examples of graphs:

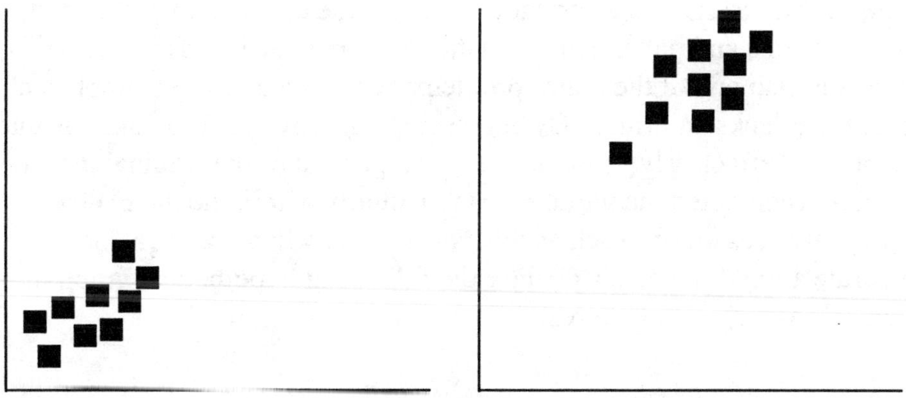

Examples of poorly scaled graphs.

Consider the graph to the left: What could be done to make it more readable? Clearly, the scales for both the independent and dependent variables are too large, and as such, include large regions of both variables' domains that do not contain data points. What makes this graph a bad one? First, the inclusion of large regions lacking data draws the viewer's attention. Second, because the scales are so large, it is difficult to see clearly any trends or patterns in the data. By recreating the graph with smaller portions of the domains of the variables plotted, one would produce a graph with significantly improved readability.

Now, consider the graph on the right above. What changes could be made to increase its readability? One simple solution would be to apply the same principles discussed above to effectively enlarge the portion of the graph that contains data. Before making such a change, one should carefully consider the implications.

Truncated scale graph.

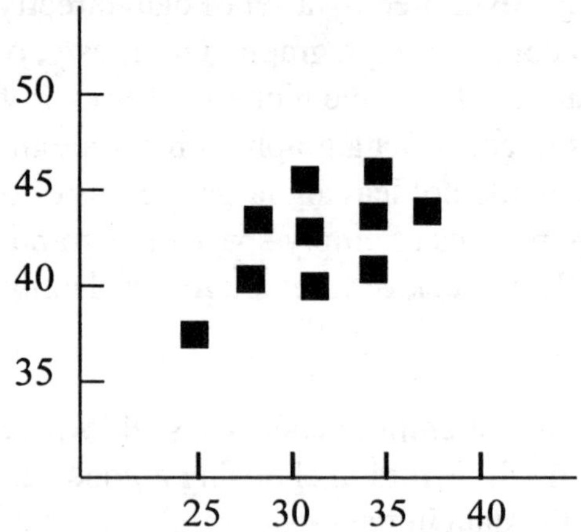

In this revised version of the graph, it is clear that the data more fully fill the field, and as such, are considerably more readable. Note that this was accomplished by plotting only a portion of the domain of each variable, and by omitting the region of the graph near the origin. At first glimpse this seems like a very good change, the data are easier to read, and variations and patterns in the data are more obvious. However, this benefit is also one of the major problems with constructing a graph such as this that lacks an origin. By decreasing the region of the domains that are plotted, effectively "zooming in" on a portion of the original graph, one has accentuated the variations and patterns within the data. One commonly sees graphs such as this constructed when one wants to illustrate a trend in data. Consider the following hypothetical graph of student's average SAT scores:

Truncated scale bar graph accenting changes in a data set.

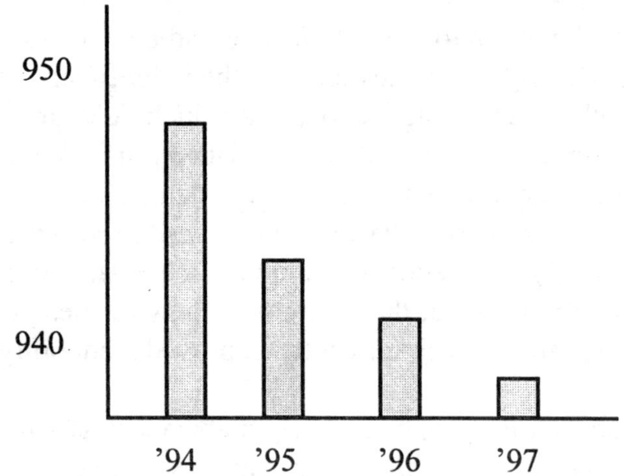

From the graph, it appears there has been a sudden and significant drop in student test scores over the past few years. However, when reconstructing the graph to include the origin (a score of zero in this case) one is able to see that while there has been a drop in scores, its magnitude is not nearly as alarming as first represented.

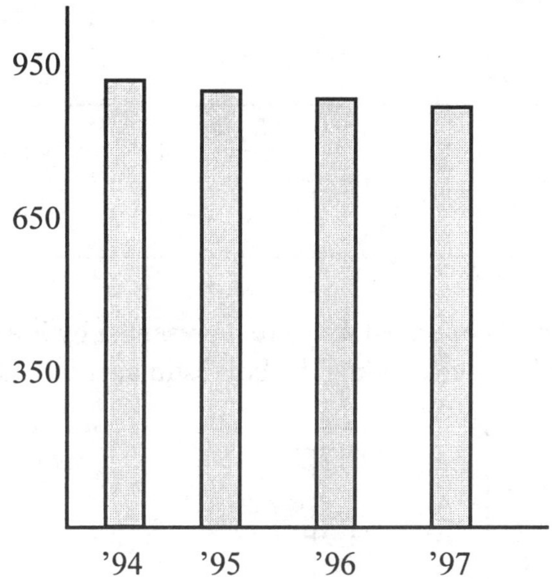

Same data plotted without a truncated scale; thereby minimizing changes.

Symbols, Data Legends, Error Bars

Once the basic framework of the graph is constructed, the data must be plotted on the field of the graph. The first consideration in this step of the graphing process is the selection of a proper symbol for the data. Commonly for a hand drawn graph, one will simply place a small dot on the graph at the designated position. However, at times one is asked to graph several sets of data on the same set of axes. In those cases, it is useful to select simple — easy to read — symbols. Suggested symbols include: circles, squares, vertical and diagonal crosses, diamonds and triangles. Occasionally it is possible to select graphical symbols which in some way serve to remind the viewer of the class of data being plotted. One runs the risk of making a graph too complicated or to "cute" when one included unusual symbols such as @, #, $, %, *, or any overly artistic or stylized symbol. The goal of symbol selection should be legibility and utility.

Selecting graphical symbols.

If more than one set of data is to be plotted on a single graph, one should include some form of data legend within or below the field of the graph to let the reader know what the symbols represent. This is particularly important if the graph is to stand largely alone without the support of explanatory text.

Data legends assist interpretation of graphs.

Finally, it is important to consider the relative uncertainty of the data that is to be plotted. In previous chapters we have discussed the nature of error in measurement and calculation. You will remember that error is commonly expressed as the range of uncertainty around a reported measure, and the actual value is expected to lie within that range. For example, one might be asked to plot the following point with associated uncertainty:

Error bars around data define a measurement's uncertainty.

x	+/-	f(x)	+/-
5	1	3	2

These levels of uncertainty are represented by bracketed lines known as error bars surrounding the best estimate of the data point (5,3).

Graph of data point and associated error bars.

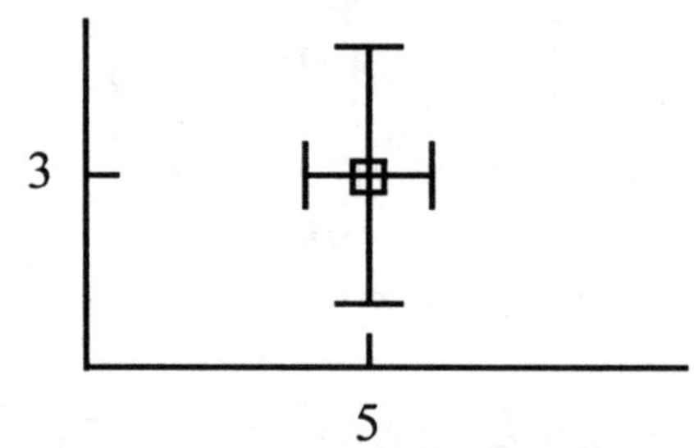

In some cases the errors associated with points in a data set are not reported, or the magnitude of the error is sufficiently small that it is ignored. In those cases, the size of the symbol is generally taken as some rough estimate of the error associated with the measure. By choosing symbols of various sizes it is possible to suggest greater continuity within data sets, or hide or accentuate trends or patterns within the data. For those reasons, one should always take care when designing or viewing symbols in order to ensure that an accurate representation of both the data point's position and its error has been given.

Chapter 5 - Exercises

Name ______________________ Date ______________________

Instructor ____________________ Section ______________________

For this exercise your instructor will supply you with a set of data that is appropriate for your particular class.

1) Which variable is independent? Which is dependent? Describe the nature of the dependence. What made you decide the dependent variable was in fact dependent?

2) Identify the maximum and minimum values of each variable, and calculate the observed range of values.

Independent Maximum ________ Dependent Maximum __________

Independent Minimum _________ Dependent Minimum ___________

Independent Range ___________ Dependent Range ______________

3) In the space below sketch a set of axes for these variables. Be sure to give consideration to all points discussed in the chapter. Should the origin of the graph be at (0,0)? Should integer scale indicators be used? Should a more complex system of scale indicators be used?

4) On the graph paper provided on page 63, again lay out your graph axes and scale indicators. Carefully plot all data on the graph. Did you select scales and axes sizes that maximize the readability and understandability of the graph? If you were to repeat the exercise, what changes would you consider making in your graph?

Chapter 6 – Statistics: Similarities and Differences

Introduction

Scientists describe nature in terms of measurements. Yet it is often impracticable or even impossible for a scientist to make measurements of all examples of the phenomenon they are studying. For instance, one may not want to measure the sizes of all the cobbles in a stream or the brightnesses of all the stars in the sky; rather, one may want to make measurements on enough of the cobbles or stars to have a fair estimate of the sizes of stones in the river or the variation in brightness of the stars. In order to deal with both the astonishing abundance and intrinsic variability displayed by natural systems, scientists have developed techniques to help them to more accurately describe nature. These techniques are generally lumped under the heading of statistics, and in this chapter we will explore some of the elementary statistical techniques used throughout the Earth sciences.

Samples and Populations

In previous chapters we have learned how scientists characterize nature in terms of numerical measurements, how they assess the reliability of those measurements, how they express changes in those measurements as functions of position or time, and how they ultimately graph and display those measurements. In this chapter we will concentrate on how scientists characterize and compare their measurements. In essence, statistics is a formalized mathematical approach to the description and analysis of large sets of data.

When we consider a set of data, the first question to be asked is: "Have all the measurements been made?" For instance, if we wanted to investigate student grades, we might ask to see all the grades of all the students. By including in our survey all grades from all students we would have the total population of all grades from which we could draw conclusions about student academic performance. Thus, the population of data is defined as all members of a data set. However, it may not be convenient or even possible to collect all the data potentially available for study. Scientists are faced with limits of time, funding for research, and even their own patience when it comes to the collection of data. Thus, it is typically the case that the analysis of some scientific problem or phenomenon results in the collection of a limited subset of all data, referred to as a sample of the population. A well-chosen sample consists of a small enough number of measures to be manageable while still large enough to be reasonably representative of the entire population. That is, analysis of the grades of a sample of all students except those with last names beginning with the letter Z might not be a significant improvement over the study of the entire population. Likewise, it might not be valid to consider the grades of only the three senior Literature majors in a class of

Research limitations force the use of statistical analysis.

Definition of a statistical population.

Definition of a statistical sample.

Greek poetry. Clearly, a meaningful and useful sample would exist somewhere between these two extremes. Proper selection of a sample for study is the first and in many cases most important step in the analysis of data.

The goal of this chapter is to give you a basic understanding of how scientists approach problems of data analysis through an illustration of how the descriptions and comparisons of samples and populations of data are conducted statistically.

Measures of Central Tendency

The analysis of a sample of data can be undertaken in many ways, and there are many interesting and important questions a scientist might wish to ask based upon that data. One of the most common types of questions dealing with data are those concerning the central tendency of the sample. Measures of central tendency are derived from all the measurements in a sample, and are as such very useful in describing that sample. There are three commonly discussed measures of central tendency; unfortunately they all have similar sounding names and are often confused one for the other — mode, mean, and median.

Calculation of the mode of a sample.

Mode – The most abundantly occurring value from a data set is defined as the mode of the sample. The mode is a statistical measure typically applied to discrete data. For instance, the most likely sum resulting from the tossing of two die is the number 7. Seven can result from combinations of 1 and 6, 2 and 5, as well as 3 and 4. No other number is as likely to occur as is 7, and it will be the modal value of any dice toss experiment. When dealing with continuous, rather than discrete, data, it is possible to define a modal class within the data. The subdivision of the continuum of observations into non-overlapping ranges will allow for calculation of the modal class of the sample.

From the set of data below, the mode is found to be 11, since that value occurs three times within the sample, at least one time more than any other value.

2, 3, 3, 4, 5, 6, 7, 8, 9, 10, 11, 11, 11, 12, 14

Calculation of the median of a sample.

Median – The median is defined as the middle value of a data set when the values of the measurements are arranged in order from smallest to largest. Thus, the number of measurements smaller than the median is equal to the number of measurements larger than the median. The median is a measure akin to those types of statistical analyses wherein a percentile is reported. That is, if you scored at the 80th percentile on a standardized test, 20% of those tested scored higher than you while 80%

scored lower. The median of a sample is defined as the 50th percentile of the sample; equal abundances of measurements are larger and smaller than that value.

From the set of data used in the calculation of the mode above, the median is found to be the number 8, since there are 7 measurements with values less than 8 (2 through 7) and 7 measurements with larger values (9 through 14).

Interestingly, for samples with an even number of data, the median is defined as the average between the two middle-most observations. Thus, in that special case, the median may be a number that is not a part of the sample! That is, for an even numbered sample of data wherein the middle two values are 6 and 7, the median would be 6.5.

Mean – The mean of a sample is the most commonly reported of all the measures of central tendency. The arithmetic mean, or average as it is more commonly called, is the sum of the values of all the data in the sample divided by the total number of data in the sample.

As we learn more about the basic statistical techniques used by scientists, we will need to begin to grasp some of the symbols that are commonly used. In terms of formal statistical symbols, the mean is defined as:

Calculation of the mean of a sample.

$$\mu = (1/n)\ \Sigma\ x_i$$

The Greek symbol μ (pronounced mu) is used to designate the mean. The symbol Σ (pronounced sigma) means: take the sum of the items that follow. The number of measurements in the sample is designated by n, and the values of each individual datum within the sample are designated by x_i.

Thus, for the set of numbers used above, the mean is calculated to be 7.73 which is the sum (116) divided by the total number of integers in the sample (15).

Measures of Dispersion

Definition of dispersion.

Dispersion within a sample or population of data is defined as the degree to which the data are spread about their central tendency. There are several ways in which to consider the dispersion within a population. We will discuss each of these in turn.

Range – The range of a sample of data is the most elementary measure of its dispersion. To calculate the range, the largest and smallest values within the sample must be identified. The difference between these values is the range.

Calculation of the range.

$$\text{Range} = x_{max} - x_{min}$$

Consider this data set:

4, 7, 9, 18, 36

In this case the range is found to be 36 - 4 = 32.

Importantly, the range of a sample only takes on some significance when it is compared to the ranges of other samples or populations. Thus, while telling us something intrinsic about the set of data at hand, it is actually more valuable as a point of comparison between data sets. Additionally, the range of a sample set is only a very crude estimator of the dispersion. It is sensitive only to the maximum and minimum values within the sample and thus fails to tell us very much about how the data are dispersed about their central tendency. That is, the range is sensitive to only the extreme values in the sample; it, however, provides no information about how the other values are distributed.

Variance – The variance is a measure of dispersion that considers the difference between the values of each datum and the mean of the sample. The variance is symbolized by σ^2 (pronounced sigma squared) and is calculated as:

Calculation of the variance.

$$\sigma^2 = \Sigma\,(x_i - \mu)^2 / (n - 1)$$

That is, the variance is calculated as the sum of the squares of the differences between each datum and the sample mean divided by one less than the number of data in the sample. Thus, it is essentially the average of the squares of the differences between the data and their mean value. The rationale for reducing the number of samples by 1 is based on the idea that the mean of the sample is not an independent value, and thus the total number of independent values is one less than the number of values in the sample. For large values of n however, the mathematical effect of subtracting 1 is obviously negligible.

What is the variance of the set of data presented in the discussion of central tendency ($\mu = 7.73$, $n = 15$)?

Example calculation of the variance of a sample.

x_i	$x_i - \mu$	$(x_i - \mu)^2$
2	-5.73	32.83
3	-4.73	22.37
3	-4.73	22.37
4	-3.73	13.91
5	-2.73	7.45
6	-1.73	2.99
7	-0.73	0.53
8	0.27	0.02
9	1.27	1.61
10	2.27	5.15
11	3.27	10.69
11	3.27	10.69
11	3.27	10.69
12	4.27	18.23
14	5.27	27.77

$$\Sigma (x_i - \mu)^2 = 187.30$$

$$\sigma^2 = 13.38$$

This series of calculations produces a value for the variance; however, this result is not directly meaningful in terms of the values of the data within the sample. In order to compare the dispersion exhibited by the sample set to the magnitudes of the values in the data (in the same units of measure) one needs to calculate the standard deviation.

Standard Deviation – This measure of dispersion is simply the square root of the variance of the sample, as noted by the symbol σ (sigma).

Calculation of the standard deviation of a sample.

$$\sigma = \sqrt{\sigma^2}$$

From the example calculation described above, the standard deviation is found to be,

$$\sigma = \sqrt{13.38} = 3.66$$

Standard deviation as a range of dispersion about the sample mean.

The standard deviation is described relative to the mean of the sample. Thus, for the example data set we have been evaluating, one would say that the mean of the sample was 7.73, and the standard deviation about the mean was 3.66. Thus the range of values within one standard deviation of the mean is defined to be bounded by ($\mu + \sigma$) and ($\mu - \sigma$) or (7.73 + 3.66) to (7.73 - 3.66) which equal the values 11.39 to 4.07.

Coefficient of Variation – Scientists are often faced with the task of comparing data from samples of various populations. For instance, one might want to know how the variability of grain sizes change in rivers from the mountains to the sea. Is the gravel and sand of high mountain streams more variable in size distribution than the sand and mud of rivers near the ocean? How would one begin to compare these two populations? The central problem with such a comparison lies in the fact that the mean grain sizes of the two samples are very different. Mountain streams are on average coarser-grained than are those near the sea. Thus, comparison of standard deviations calculated from samples derived from either population would be difficult. In order to facilitate comparisons of the variance in populations of data with different means, scientists have developed the notion of the coefficient of variance.

Calculation of the coefficient of variation.

The coefficient of variance is defined as 100 times the standard deviation divided by the mean.

$$CV = (100\ \sigma) / \mu$$

Thus, samples of data that exhibit large deviations relative to the value of their mean have large coefficients of variance, while those with standard deviations that are small relative to their mean are found to display small coefficients of variance. From the example given for particles in stream beds, one might find that the standard deviation (as measured in millimeters) for sediment from mountain streams is very large relative to that of more distal rivers while the coefficients of variance calculated from both samples might in fact be very similar.

Samples as Predictors of Populations

Collecting a representative sample.

A scientist must always remember that the collection of a sample represents only a portion of the total population. While samples collected and measured for statistical analysis are often very large, they still commonly represent only a tiny fraction of the total population. As such, one is always faced with the necessity of establishing how well a particular sample serves as a predictor of the population of interest. The first step to generating a good sample lies in the design of the procedure of data collection. It is important to ensure that the process of obtaining data does as little as possible to alter (or more properly stated — bias) the samples relative to the population. For instance, one would not want to set out to analyze the sizes of a collection of rock samples from a mountain steam by using a teaspoon. Obviously, it would be very difficult to successfully acquire samples much larger than the spoon. Thus, one would bias the distribution of sizes collected relative to the true population. That example seems trivial and obvious, but there are many cases in the history of science where biased measurements were made, and erroneous conclusions then were drawn from poorly sampled populations.

Of particular interest to scientists is the likelihood that a measured sample mean is a valid estimate of the mean of the entire population. Unfortunately, statistics is unable to give us an exact answer to that question. Rather, it does provide a measure of the likelihood with which we can be sure that the mean of the population is near the mean of the sample. In order to address this problem adequately, one must specify the level of uncertainty one is willing to accept in such an estimate. Are you willing to be wrong 10% of the time? Only 5% of the time? Less than 1% of the time? The acceptable level of uncertainty is dependent upon the application in question, and the ease with which data can be collected. The more data available in the sample, typically the smaller the range of uncertainty.

Establishing a level of certainty.

The likelihood that the mean of the sample is close to the mean of the entire population is expressed in terms of a confidence interval and is calculated as

Calculation of a confidence interval about the sample mean.

$$CI = \mu \pm (K\,\sigma) / \sqrt{n}$$

where K is a statistical constant reflecting the confidence one is requiring of the estimate.

If a situation demands that there be great confidence that the mean of the population lies within a given range around the sample mean then it is necessary for the range around the sample mean to be larger than if one only demands a lesser certainty.

Table of confidence interval constants.

Confidence	***Constant (K)***
90%	1.58
95%	1.96
99%	2.55

Examples of confidence interval calculation.

Consider an example wherein the mean of a sample is 10, the standard deviation is 2, and the number of values in the sample is 100.

90% Confidence Interval About the Sample Mean

$$CI(90\%) = 10 \pm (1.58 \times 2) / \sqrt{100}$$

$$CI(90\%) = 10 \pm 0.32$$

$$CI(90\%) = 9.68 \text{ to } 10.32$$

95% Confidence Interval About the Sample Mean

$$CI(95\%) = 10 \pm (1.96 \times 2) / \sqrt{100}$$

$$CI(95\%) = 10 \pm 0.39$$

$$CI(95\%) = 9.61 \text{ to } 10.39$$

99% Confidence Interval About the Sample Mean

$$CI(99\%) = 10 \pm (2.55 \times 2) / \sqrt{100}$$

$$CI(99\%) = 10 \pm 0.51$$

$$CI(99\%) = 9.49 \text{ to } 10.51$$

Thus, one is able to conclude with greater confidence that the population mean lies within a given range of the sample mean as the size of that range around the mean is increased.

Evaluating Linear Relationships

One of the basic tasks scientists set themselves to is the discovery and analysis of patterns within data. Perhaps one of the most common and important types of pattern is that of the linear relationship. Linear relationships are typically discussed with regard to the way in which a dependent variable changes as a function of an independent variable. When a scientist describes and discusses linear relationships there are two important aspects they consider: the degree to which there exists a linear relationship; and the equation of the line that best describes that relationship.

Correlation Coefficient – When two sets of data are plotted against each other on a graph, one of four sorts of results is typically found:

Examples of correlation between variables.

No relationship (Random)

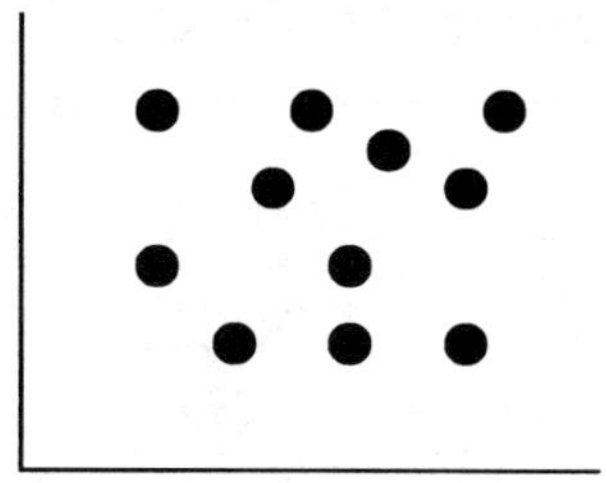

Independent Relationships (One Variable Invariant)

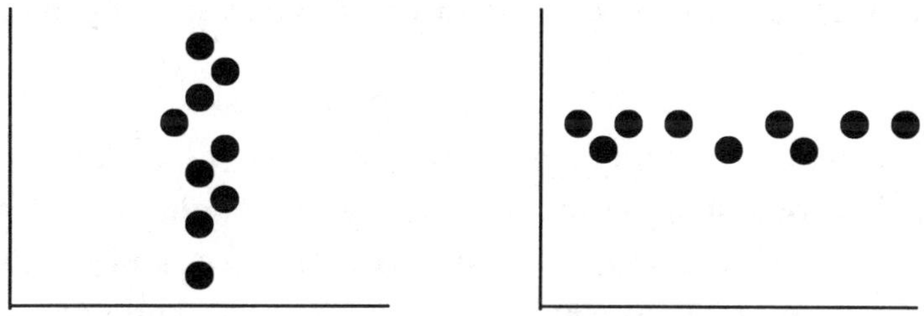

Weakly Linear Relationship

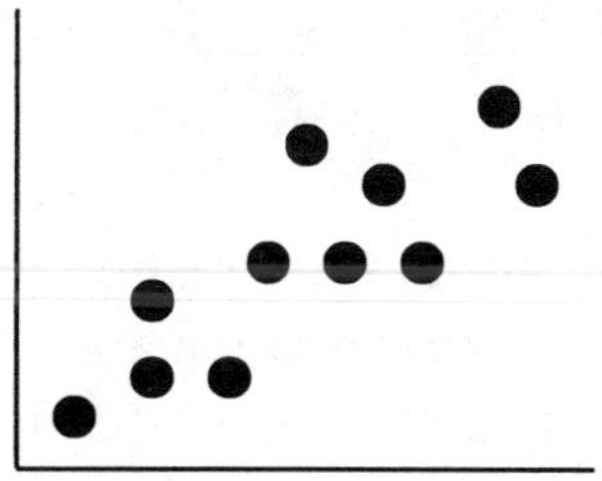

Perfect Linear Relationship

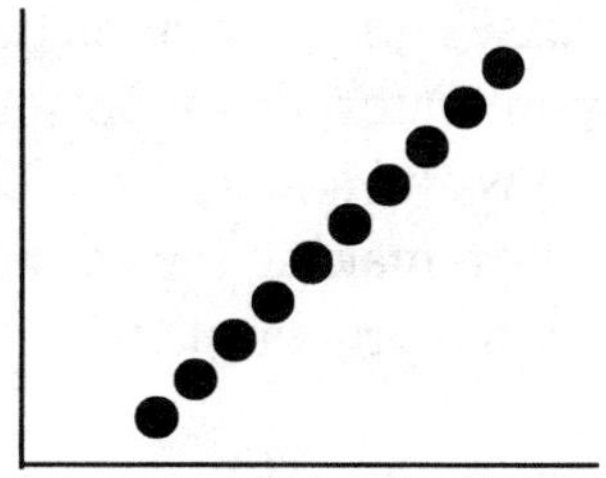

The degree to which a scientist's data describe one of these various types of relationships is a measure of the correlation between the variables. In Chapter 4 we discussed correlation in general terms, now we will begin to quantify it. The statistical measure of correlation commonly used is the correlation coefficient. In symbolic form, the equation for the calculation of correlation coefficient is a bit daunting.

Calculation of the correlation coefficient.

$$r = \frac{\Sigma(XY) - \frac{(\Sigma X)(\Sigma Y)}{N}}{\sqrt{\left[\Sigma(X^2) - \frac{(\Sigma X)^2}{N}\right]\left[\Sigma(Y^2) - \frac{(\Sigma Y)^2}{N}\right]}}$$

However, by breaking this equation down into its various parts it is easily understood.

Explanation of summations.

The fundamental mathematical procedure in the calculation of the correlation coefficient is the summation (Σ). Throughout the equation there are several different types of summations. Consider the two samples of data we wish to compare, X and Y. In the numerator of the equation is the symbol $\Sigma(XY)$ that means: multiply each pair of values of X and Y together, and sum those products for all values. The other two summations in the numerator are simply the sums of all the values of X (ΣX) and all the values of Y (ΣY) that are then multiplied together and divided by the total number of pairs of data.

The entire denominator of the equation lies under the square root sign, which indicates that all other mathematical operations are carried out before the square root is taken. Under the square root there are several summations to be evaluated. The first type is symbolized as $\Sigma(X^2)$ and $\Sigma(Y^2)$. These symbols indicate that for each value of X and Y, one is to first square that value (multiply with itself) and then add up all the squared values. This is a different operation than those symbolized by $(\Sigma X)^2$ and $(\Sigma Y)^2$. In these cases one first sums the individual values of X and Y as was done in the numerator, and then squares the resulting sum. It is these various summations and squares that are most easily confused. One must be very careful when making the calculations to ensure the proper value is positioned at the proper place within the equation.

Evaluating the correlation coefficient.

The correlation coefficient *r* takes on values between 1 and -1. A value of positive one indicates perfect linear correlation with a positive slope, while a value of negative one indicates a perfect linear correlation with a negative slope (see Chapter 4). A value of zero indicates no correlation. Generally speaking, the closer the value of the coefficient is to either 1 or -1 the stronger the correlation between the variables.

However, one should always remember that the presence of a correlation between variables only implies, it does not demand, that there be a causal relationship between the variables.

Finding Linear Equations – Establishing the presence of a linear relationship between two variables is only the first step in their analysis. Beyond the simple calculation of the correlation coefficient a scientist would also like to define the equation for the line that best fits the data. You will recall that the equation for a line takes on the form:

Equation of a line.

$$Y = mX + b$$

Where X and Y are the independent and dependent variables respectively, m is the slope of the line, and b is the location of the intersection of the line with the Y-axis (the so-called Y-intercept). How are the constants *m* and *b* calculated from a set of samples?

The equation for the slope (*m*) is

Calculation of the slope of a line.

$$m = \frac{\Sigma(XY) - \dfrac{\Sigma(X)\Sigma(Y)}{N}}{\Sigma(X^2) - \dfrac{(\Sigma X)^2}{N}}$$

Again in this case one needs to be sure to calculate carefully the proper summations in order to get the correct result.

The equation for the Y-intercept is much easier:

Calculation of the Y-intercept of a line.

$$b = \mu_Y - m\mu_X$$

That is, the value of the intercept is the slope (as calculated above) multiplied by the mean of the X values, subtracted from the mean of the Y values.

Chapter 6 Exercises

Name ______________________ Date ______________________

Instructor ______________________ Section ______________________

Carry out the following statistical calculations using the data table provided on page 85.

1) The variable X represents a continuous variable. In order to calculate a mode one must subdivide the range of X into equal sized intervals known as bins. For our purposes, each bin should cover 5 units. That is, the first bin goes from 0.01 to 5.00, the second from 5.01 to 10.00, the third from 10.01 to 15.00, and so on.

Bin	Frequency
0.01 - 5.00	_______
5.01 - 10.00	_______
10.01 - 15.00	_______
15.01 - 20.00	_______
20.01 - 25.00	_______
25.01 - 30.00	_______

Modal Class of X = _______

2) Find the median value of X; note that the values of X are arranged in order of increasing magnitude.

Median of X = _________

3) Calculate the mean of X; note that there is also a place for you to record this value directly below the data table.

Mean of X = _________

4) Calculate the range of X.

Maximum value of X __________

Minimum value of X __________

Range of X = ___________

5) Calculate the variance and standard deviation of X; show all work.

Variance of X = ____________

Standard Deviation of X = _______________

6) Calculate the variance and standard deviation of Y; show all work.

Variance of Y = _______________

Standard Deviation of Y = _____________

7) Calculate the coefficients of variation of both X and Y; show all work.

CV_X = ________

CV_Y = _________

Which is larger? Briefly discuss why.

8) Using the statistical constants given in chapter 6, calculate the 90, 95, and 99% confidence intervals around the mean of X; show all work.

90% CI_X = ________

95% CI_X = __________

99% CI_X = ___________

9) Using the graph paper provided construct a graph using the values of X as the independent variable, and Y as the dependent variable. Label and scale axes appropriately.

10) Calculate the correlation coefficient between the values of Y and X; show all work.

Correlation Coefficient = ______________

11) Calculate the slope of the best-fit line describing the linear relationship between X and Y; show all work.

Slope (m) = ____________

12) Calculate the Y-intercept values for the best-fit line describing the linear relationship between X and Y; show all work.

Y-intercept (b) = ____________

13) Write the equation of the best-fit line describing the linear relationship between X and Y; plot this line on the graph.

Equation of the line ________________________________

Reference	X	$X-\mu_X$	$(X-\mu_X)^2$	X^2	Y	$Y-\mu_Y$	$(Y-\mu_Y)^2$	Y^2	XY
1	1.04				14.22				
2	2.74				6.97				
3	5.54				13.69				
4	6.17				18.61				
5	7.35				27.56				
6	7.36				15.79				
7	8.00				11.22				
8	11.30				16.76				
9	11.58				21.09				
10	12.59				20.74				
11	13.97				34.04				
12	14.07				25.17				
13	15.45				32.81				
14	15.82				27.31				
15	17.15				41.90				
16	17.27				30.93				
17	18.27				37.73				
18	19.08				44.24				
19	19.25				36.44				
20	20.84				52.07				
21	22.02				56.28				
22	22.27				56.56				
23	25.44				57.29				
24	25.58				49.33				
25	25.79				62.70				
26	25.85				49.92				
27	25.89				55.40				
28	26.36				48.52				
29	27.02				67.62				
30	27.93				66.30				
31	29.41				66.67				

ΣX __________ $(\Sigma X)^2$ __________ $\Sigma(X-\mu_X)^2$ __________ ΣX^2 __________

ΣY __________ $(\Sigma Y)^2$ __________ $\Sigma(Y-\mu_Y)^2$ __________ ΣY^2 __________

$\Sigma(XY)$ __________

Chapter 7 – Logarithms: Calculation and Application

Introduction

Natural systems are dominated by extreme values; as we learned in Chapter 1 numbers that range from extremely large to extremely small are common. Early in the history of science it was recognized that calculations involving either very large or small numbers (specifically multiplication, division, and trigonometric functions) were both difficult and commonly subject to human error. To facilitate mathematical manipulations of these types, a system of calculation known as the logarithms (ratio-numbers) was developed early in the 17th century. Since then, systems displaying logarithmic relationships have been identified and described from all branches of the physical and natural sciences. In this chapter we will explore the basics of logarithmic calculation and discuss some of the more important applications of logarithms to the Earth sciences.

Historical Background

With the advent of telescopic observation in the late 16th century, data generated by astronomers became both readily abundant and precise. However, this rapid expansion in the quantity and quality of observational data posed a serious problem to scientists. In the days before hand-held calculators and digital computers it was necessary to carry out all mathematical calculations by hand. In order to fully utilize all the astronomical data that had become available, scientists needed a shortcut method to speed up such simple mathematical manipulations as multiplication, division, calculation of square roots, as well as calculation of the various trigonometric functions.

Origin of logarithms.

In 1614, the Scotsman John Napier published the first table of logarithms under the somewhat fanciful title *Description of the Wonderful Law of Logarithms*. Napier's original work was subsequently expanded and improved by Henry Briggs and Joost Burgi in the 1620s. Since that time logarithms have become an integral part of science and mathematics.

Logarithmic calculations are a central part of the Earth sciences. After introducing logarithmic forms and their calculation, we will explore the application of logarithms to the study of earthquakes, sediment grain sizes, and the acidity of natural solutions.

Logarithmic Forms

Definition of a logarithmic function.

The fundamental definition of a logarithm is given by the equation:

$$x = (a)^{\log_a(x)}$$

Definition of a logarithmic base.

Thus, the logarithm of the number X is the power to which a base (a) must be raised to return a result of X. The base must be a positive number greater than 1. Three general rules of logarithms follow when $a > 1$.

Rules of logarithms.

Rules of Logarithms

$$\log x > 0 \text{ if } x > 1$$

$$\log x < 0 \text{ if } 0 < x < 1$$

$$\log x = 0 \text{ if } x = 1$$

The most general form of the logarithm is written as $\log_a x$ which is read "the base a logarithm of x," where x can take on any value as above, and a must have a value greater than 1.

The base 10 logarithmic system.

Throughout the sciences one of the most useful and widely used logarithmic bases is base 10, known as the common logarithm. The power and beauty of the base 10 logarithm lies in its utility to express powers of 10 such as those described in our discussion of scientific notation in Chapter 1. The common logarithm has the property that it returns the value of the exponent to which the number 10 is raised. For example:

Relations between logarithms and exponents in the base 10 system.

$$\log 0.00001 = \log 10^{-5} = -5$$
$$\log 0.0001 = \log 10^{-4} = -4$$
$$\log 0.001 = \log 10^{-3} = -3$$
$$\log 0.01 = \log 10^{-2} = -2$$
$$\log 0.1 = \log 10^{-1} = -1$$
$$\log 1 = \log 10^{0} = 0$$
$$\log 10 = \log 10^{1} = 1$$
$$\log 100 = \log 10^{2} = 2$$
$$\log 1000 = \log 10^{3} = 3$$
$$\log 10000 = \log 10^{4} = 4$$
$$\log 100000 = \log 10^{5} = 5$$

Clearly, calculation of the common logarithms is straight forward for the simple powers of 10; however, how does one find the base 10 logarithm of a more elaborate number such as 54.693? The first and most important characteristic of logarithmic functions is that they are non-linear. Consider the graph of the common logarithms of numbers between 1 and 10.

Graph of the base 10 logarithms of the numbers 1 through 10.

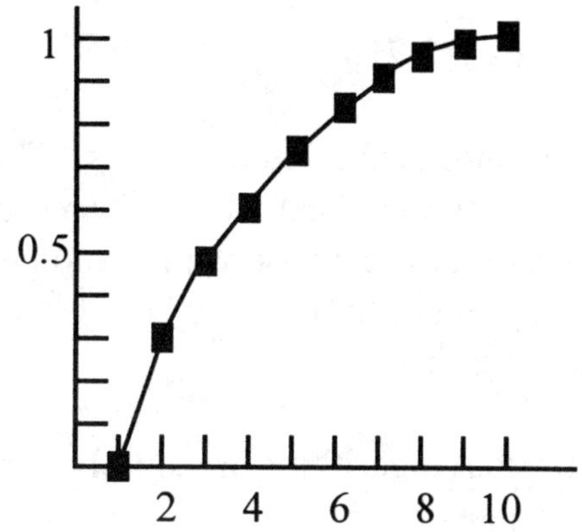

Procedure for calculating a logarithm.

How does one use this relationship to go about finding the common logarithm of a number such as the one given above (54.693)? The first step is to rewrite the number in scientific notation:

$$54.693 = 5.4693 \times 10^1$$

From there, one is able to utilize the basic law of logarithms:

$$\log xy = \log x + \log y$$

which means that by taking the logarithm, one is able to alter the expression of a number to a simple form such as:

$$\log 54.693 = \log 5.4693 \times 10^1$$

$$\log 5.4693 \times 10^1 = \log 5.4693 + \log 10^1$$

$$\log 5.4693 + \log 10^1 = \log 5.4693 + 1$$

and by inspecting the graph above, one can see that

$$\log 54.693 = 1 + \log 5.4693 \simeq 1 + 0.74 \simeq 1.74$$

Definitions of the characteristic and mantissa of a logarithm.

The integer part of the answer, reflecting the power of 10 to which the original variable was raised, is known as the characteristic. The decimal fraction derived from the calculation of the common logarithm for values between 1 and 10 is known as the mantissa.

The base e logarithms.

There are two other important logarithmic bases commonly used throughout the sciences: base 2 and base e. The base e, or natural logarithms, are used in such diverse fields of study as biological growth and the decay of radioactive elements. The constant e, the natural exponential function, is an irrational number approximated as

$$e \sim= 2.718$$

and is one of the truly important numbers in mathematics. The source of the significance of e lies in its utility in the calculus; otherwise, there is no particular advantage to the selection of one base versus another in most scientific applications. Typically, logarithms to the base e are designated by the symbol ln(x) rather than $\log_e (x)$.

It is important to realize that logarithms are simply exponents, and as such follow the laws of exponents.

Laws of logarithms.

Laws of Logarithms

$$\log xy = \log x + \log y$$

$$\log x/y = \log x - \log y$$

$$\log x^a = a \log x$$

$$\log \sqrt[a]{x} = (1/a) \log x$$

Changing the Base

At times it is necessary to change the base of a logarithm; typically one finds cause to change bases between the common and natural logarithm systems. Given a logarithmic equation in one base, the following steps allow for transformation to another:

Calculating a change in logarithm base.

$$a^{\log a (y)} = y$$

Take the logarithm of both sides utilizing the third rule given above,

$$[\log_a (y)] \times [\log_b (a)] = \log_b (y)$$

$$\log_a (y) = \log_b (y) / \log_b (a)$$

and since the function $\log_b(a)$ is a constant, it follows that logarithms of all bases are related to each other by constant ratios. For instance, to change the expression of a logarithm from base 10 to base e, the following formulation is used:

Changing from base 10 to base e.

$$\log_e (x) = (2.302585) \log_{10} (x)$$

Similar kinds of equations can be derived for reversing the change of base, or switching to another base altogether.

Logarithms in the Earth Sciences

Logarithmic earthquake magnitudes.

One of the most well-known examples of logarithms in the Earth sciences is drawn from the field of geophysics. A simple measure of the energy released during an earthquake is given by the magnitude of the event. However, because earthquakes are found to occur over a very wide range of intensities, the magnitudes of the earthquakes are calculated according to a logarithmic scale. The base 10, or common logarithm, provides the structure for describing the energy release (and thus the potential damage that ensues) during an earthquake. Major earthquakes commonly range in magnitude from 5 to 8. What do these numbers mean?

Earthquake magnitude is reported according to a base 10 scale; therefore, an increase in the magnitude by one unit represents a ten-fold increase in the energy released. That is, a magnitude 6 earthquake is ten times more powerful than a magnitude 5 event. Commonly however the magnitudes of earthquakes are reported to at least two significant figures. How much more powerful is a 6.7 earthquake relative to a 5.2? To calculate this ratio, consider the following:

Evaluating the relative strengths of two earthquakes.

$$\text{Energy} \sim= 10^M$$

$$E_1 = 10^{6.7}$$
$$E_2 = 10^{5.2}$$

$$E_1/E_2 = 10^{6.7} / 10^{5.2}$$

$$E_1/E_2 = 10^{(6.7-5.2)}$$

$$E_1/E_2 = 10^{1.5}$$

$$E_1/E_2 \sim= 32$$

Thus, the magnitude 6.7 earthquake reflects an approximately 32 times greater release of energy than does a magnitude 5.2 earthquake!

Logarithmic grain sizes.

A second important use of the logarithmic scale in the Earth sciences comes from the study of particle sizes. Sediment in streams and rivers is found to exist over a wide range of sizes from boulders to silt and mud. In order to erode, transport, and deposit sediment a flowing current is needed. Scientists studying the movement of particles by currents have found that there is a non-linear relationship between the diameter of a sedimentary particle and the current energy needed to transport that grain, and have therefore chosen a logarithmic scale to represent grain diameter.

There are some interesting aspects of the logarithmic scale as used by sedimentologists. The first is that it is a base 2 system, and the second is that the value of the logarithm is multiplied by negative one, such that grains smaller than 1 millimeter in diameter are represented by positive values and grains larger than 1 millimeter by negative values. The symbol chosen to represent this logarithmic grain scale is ϕ (pronounced phi).

The base 2 grain size scale.

$$\phi = -\log_2 d$$

And as such, the following table describes grain sizes in terms of ϕ units.

Table of grain diameters.

Diameter (mm)	***Diameter*** ϕ
64	-6
32	-5
16	-4
8	-3
4	-2
2	-1
1	0
1/2	1
1/4	2
1/8	3
1/16	4
1/32	5
1/64	6

A second important negative logarithmic system is one that most everyone is generally familiar with, the pH scale. The acidity of a water is defined as:

$$pH = -\log_{10} [H^+]$$

The negative base 10 logarithmic scale of acidity.

where H is the concentration of hydrogen atoms (or more correctly, the chemical activity of those atoms) as measured in moles per liter. In this case, the concentration of hydrogen atoms is typically less than one mole per liter, and by multiplying by negative one, the resultant value is positive, and thus a more convenient unit of measure. Neutral water, neither acidic or basic, has a pH of 7 reflecting a concentration of 1×10^{-7} mol/L [H+]. Acidic water has a pH value less than 7 meaning more H^+, while basic water has a pH values greater than 7 meaning less H^+.

These three examples illustrate the variety of applications to which logarithmic functions have been put. Any system that exhibits variation over large ranges is often best characterized through the use of logarithmic analysis.

Chapter 7 - Exercises

Name ______________________ Date ________________________

Instructor ___________________ Section ______________________

1) Use your understanding of logarithms and the graph of the common logarithm function from page 85 to estimate the following logarithmic transformations. Show all work.

a) $\log_{10} (1/467)$

b) $\log_{10} 687$

c) $\log_{10} 5946.32$

d) $\log_{10} (1/1000)$

e) $\log_{10} 2.41$

f) $\log_{10} (1/64)$

2) Use a pocket calculator to check your answers from Question 1.

a) $\log_{10} (1/467)$

b) $\log_{10} 687$

c) $\log_{10} 5946.32$

d) $\log_{10} (1/1000)$

e) $\log_{10} 2.41$

f) $\log_{10} (1/64)$

3) Solve the following natural (base e) logarithms using a pocket calculator.

a) $\ln 16$

b) $\ln 0.865$

c) $\ln 0.0016$

d) $\ln 97$

4) Convert the following common logarithms to base e, and then base 2 logarithms. Show all work.

a) $\log_{10} 1000$

b) $\log_{10} (1/417)$

c) $\log_{10} 4.669$

d) $\log_{10} 0.00047$

5) Estimate the increase in energy between a magnitude 4.7 and a magnitude 2.4 earthquake. Show all work.

6) What is the difference in diameter in millimeters between a 3.54 ϕ and a 1.69 ϕ grain? Show all work.

7) Express in terms of scientific notation the difference in concentration (mol/L) between a solution with a pH of 4.5 and a solution with a pH of 6.9. Show all work.

Chapter 8 – Symmetry: Shape and Pattern

Introduction

Nature abounds with symmetry . Everywhere we look we observe the organization of the complexity of natural systems into easily recognized shapes and patterns. Above us the clouds arrange themselves into long parallel bands spaced at equal intervals across the sky. Below us mineral crystals display beautiful and complex three-dimensional shapes. The most ubiquitous and important type of spatial organization displayed by nature is known as symmetry. Symmetries exhibited by natural systems come in many different types. In this chapter we will explore the varieties of symmetry, and from those descriptions develop a basic understanding of the underlying origins of such patterns.

Simple Symmetry

To be symmetrical, an object must have parts or regions that are exactly "transformable" into its other parts or regions. This notion of transformation is a bit slippery. Essentially, what is meant by the word transformation is a change in an object wherein everything stays the same. That is, after a symmetry transformation, the object is found to have exactly the same shape or pattern that it exhibited prior to the transformation. The type of symmetry with which we are most intimately familiar is termed bilateral symmetry. Roughly speaking, the human form is symmetrical about an imaginary plane that separates the left-hand from the right-hand side. Two eyes, two ears, equal distributions of teeth and toes, fingers and feet, on either side of this imaginary plane. However, the vertical plane of symmetry that divides the body into two essentially identical parts has no other counterparts within the human form. Dividing horizontally clearly demonstrates that the upper half of our body is very dissimilar from the lower half. Likewise, division into front and back halves also indicates a lack of symmetry between those regions. Thus, while symmetry is amazingly abundant in nature it is clearly not universal. Interestingly, in an aesthetic sense, nearly perfect symmetry is commonly considered to be more attractive than mathematically perfect symmetry.

Definition of symmetry transformations.

In order to understand the symmetry of nature, and ultimately its underlying causes, we must undertake a description of the types of symmetry transformations possible. In turn, we will discuss translation, reflection, rotation, and rotoinversion.

The four types of symmetry transformations.

Translational Symmetry

Symmetry of translation is the most elementary of all the transformation operations. One-dimensional translation results in a linear pattern wherein the original shape is repeated at a constant interval along the line of translation. This spacing interval determines the geometry of the translational symmetry.

Translation along a line.

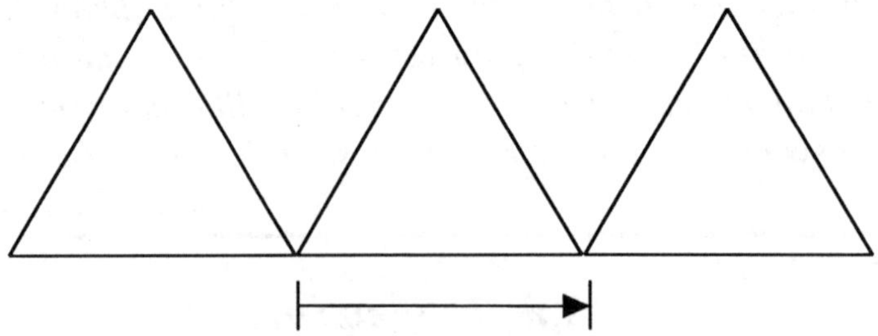

Translation Interval

Translations are not limited to one-dimensional linear patterns. Two and three-dimensional translational symmetries are common in nature. Typically the patterns formed by translational symmetry are described as tillings, because they fill an area similar to the way in which ceramic tiles cover a floor or wall. Importantly, a large number of shapes can be arranged so that the entire area is covered. However, there are an even greater number of shapes that cannot be arranged to cover the entire area.

Geometric shapes that tile area without gaps or overlaps.

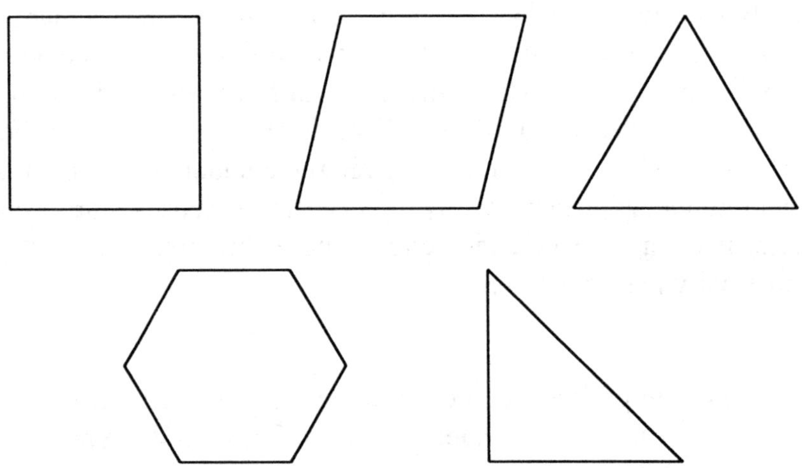

The shapes above can be easily arranged so that the repetition of any individual geometry across a planar surface will cover all the area with no gaps or overlaps.

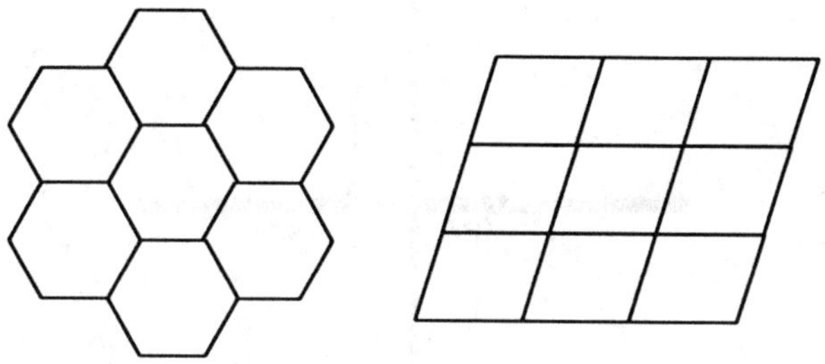

Examples of two-dimensional tilling.

The patterns generated by the tilling of those shapes clearly differs from that of the shapes below, the repetition of which will not cover all area without either overlaps or gaps between the forms.

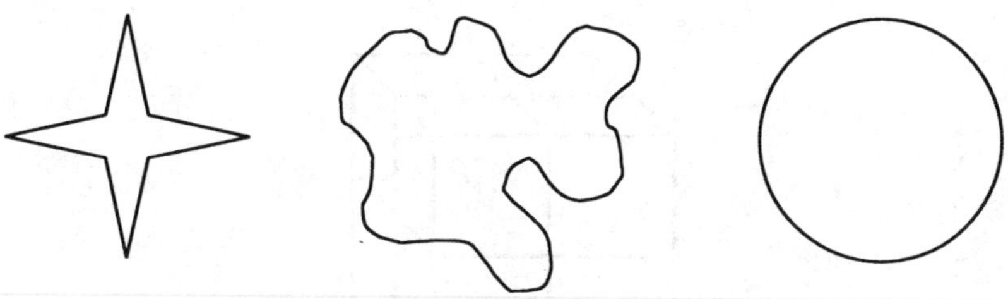

Examples of shapes that will not tile without gaps or overlaps.

Reflectional Symmetry

The symmetry of reflection represents another elementary form of symmetry transformation. The bilateral symmetry discussed above serves as a simple example of symmetry by reflection. On a two-dimensional plane, such as this piece of paper, regions of an object that are transformable through a reflectional transformation are divided from each other by a line of symmetry.

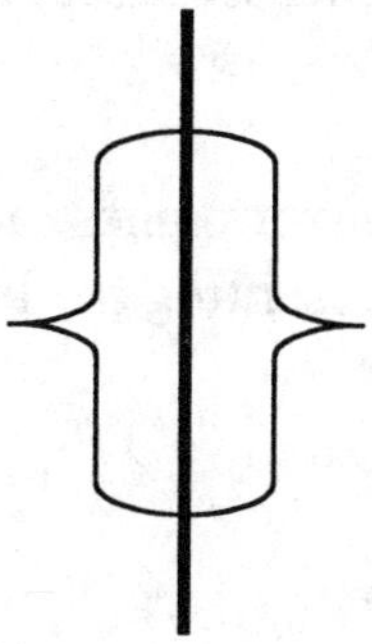

Example of reflectional symmetry.

Discovery of a second line of symmetry.

Simply by "flipping" the left-hand side of the object over the line of symmetry it is possible to form an exact replica of the right-hand side. The careful observer will also notice that the particular objects drawn above are also symmetrical about a horizontal line.

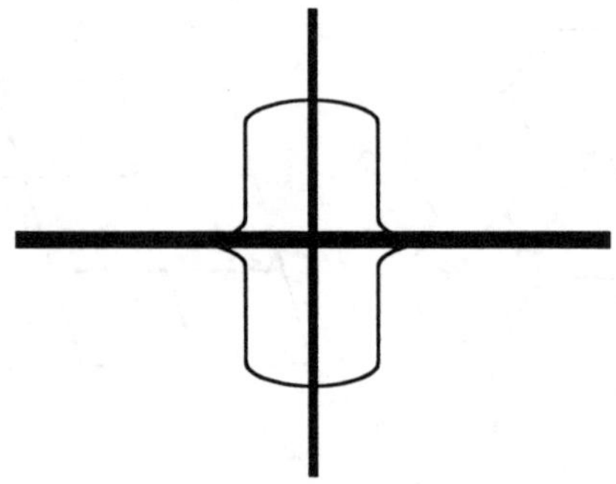

In a three-dimensional object, regions translatable by a reflection are divided by planar surfaces, known as mirror planes. A simple shape such as a cube is divided by three surfaces representing the single horizontal and two vertical (left-right and front-back) mirror planes, each of which equally and symmetrically divides the cube. Taken together the original large cube is divided by the three mirror planes into eight smaller cubes.

A cube has three mirror planes.

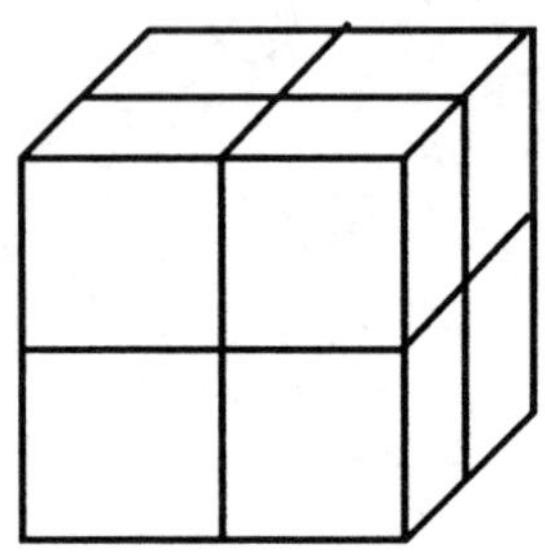

Rotational Symmetry

Symmetry of rotation is closely related to reflectional symmetry. The nature of the rotational symmetry within an object defines the number of mirror planes present. Consider the following simple examples in two-dimensions.

A rectangle has two mirror planes (actually mirror lines in two-dimensions) of symmetry representing the long and the short dimensions of the form respectively.

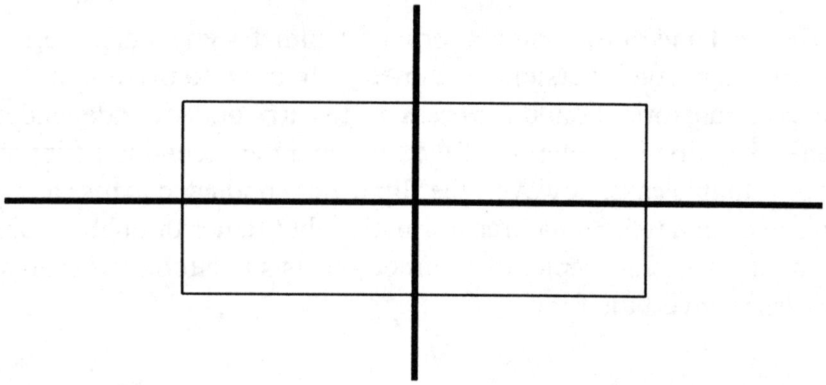

The intersection of two mirror planes defines a rotational axis.

The intersection of these two mirror planes defines an axis of rotational symmetry that runs perpendicular to the plane of this page (into and out of the page). Around this axis, the rectangle can be rotated. A rotation of 180° clockwise or counterclockwise will result in the shape becoming transformed perfectly on top of itself. A second rotation of 180° will likewise result in a perfect transformation. A rotation of any more or less than 180° will cause a misalignment of the object and thus a imperfect transformation. This kind of rotational symmetry is known as transformation about a two-fold rotational axis. What would a three-fold, four-fold or even eight-fold rotational symmetry look like?

Rotational symmetry is easily extended to three-dimensions. As shown above, a cube has three mirror planes intersecting at right angles that combine to form the three axes of rotation. The intersection of each pair of mirror planes defines a line in space. The horizontal and left-right vertical mirror planes define a front-to-back axis. The horizontal and front-back vertical mirror planes define a left-to-right axis. While the two vertical mirror planes define a top to bottom axis.

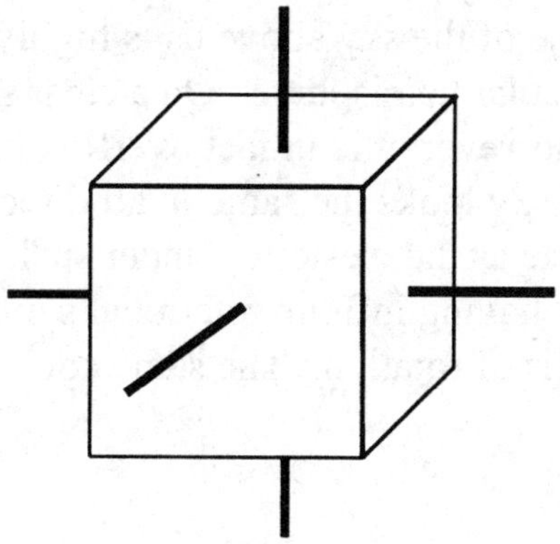

Axes of rotation defined by intersection of mirror planes in a cube.

Rotoinversional Symmetry

The final and most complex form of symmetry encountered in nature is known as rotoinversional symmetry. In order to perform a rotoinversion transformation, it is necessary to carry out two independent operations. The first is a rotation of the form and the second is a flipping of the form. In this case, however, the flip is not conducted across a mirror plane as in a refectional transformation, but rather through a point. This point, known as the center of symmetry, acts something like a lens in that it serves to invert the form.

Inversion through a center of symmetry.

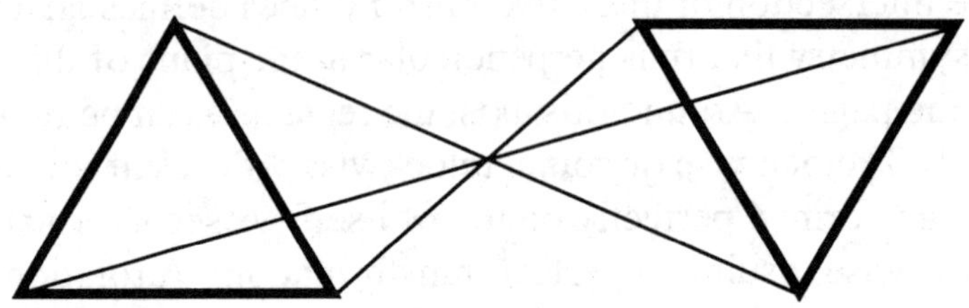

The center of symmetry, or lens, is defined to be the point through which lines connecting the flipped regions of each shape intersect. Rotoinversion transformations are important in three-dimensional objects, but prove to be rather difficult to describe in two dimensions. Several common three-dimensional objects such as cubes and octahedrons have important rotoinversional symmetries.

Reduction of Symmetry and the Appearance of Pattern

Nature abounds with symmetry. However, most of that symmetry goes unnoticed. The dome of the sky above us is highly symmetrical. It can be thought of as a circular hemisphere. On a clear sunny day, the sky appears to be uniform. However, it is in fact overflowing with symmetries. Such a blue sky looks the same in all directions, the eastern sky is just as clear and blue as the western. Under such conditions, the sky could be described as having infinite rotational symmetry with your position serving as the axis of rotation. The same could be said for a perfect circle or sphere.

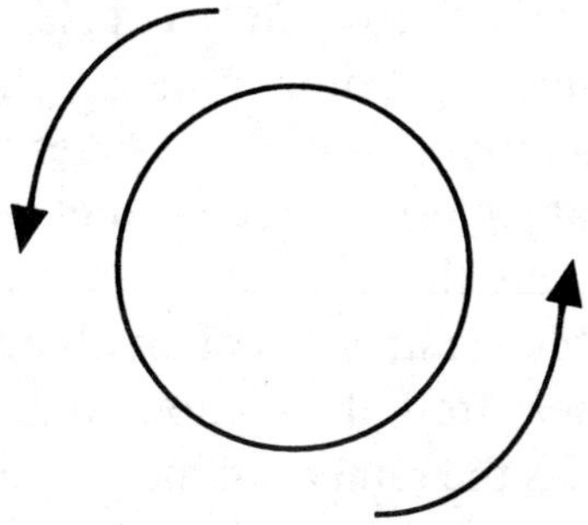

Infinite rotational symmetry of a circle.

A rotation of any arbitrary angle about the central axis will result in the circle's form remaining unchanged. Consider however the case of a sphere. By extending the shape from two-dimensions (circle) to three-dimensions (sphere) the number of axes about which any rotational transformation can occur has increased from 1 (circle) to an infinite number (sphere). The sphere is the single most symmetrical three-dimensional geometry possible.

This infinite symmetry can be reduced, or broken, by painting a circular line around the sphere's equator. The presence of this line, somewhat like the difference between the cue ball and a striped billiard ball, greatly reduces the symmetries present within the form. The uniformly white cue ball is infinitely symmetrical. The striped ball has been reduced to only two axes of rotational symmetry.

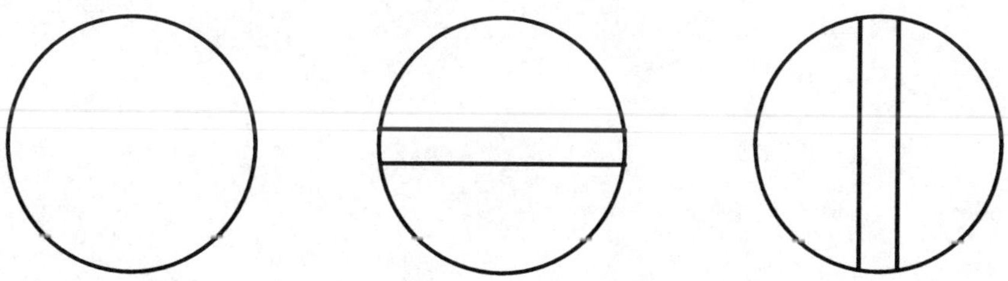

Cue Ball ***Stripe Along Horizontal*** ***Stripe Along Vertical***

Symmetry breaking in a sphere.

When the stripe is positioned along the horizontal equator of the ball, the sphere has an infinite rotational symmetry about a vertical axis. When the stripe is placed along a vertical plane, the vertical axis of rotation is reduced from an infinite number of equivalent positions to a simple two-fold rotational symmetry.

Symmetry breaking produces pattern.

What is the point to this discussion? It is clear that uniform geometries have a lot of symmetry, while more irregular objects have less. However, the most important point to consider from this discussion of symmetry breaking is that patterns, textures, and designs which we recognize throughout nature are the result of a reduction of symmetry. Only by reducing the possible symmetries of an object does a pattern such as a striped billiard ball appear from the monotonous white sphere of the cue ball. Thus, symmetry has the rather unique property of being most obvious when it is present in the most sparing degree. The less symmetrical an object or pattern, the more obvious the symmetry to the eye.

Chapter 8 - Exercises

Name ______________________ Date ______________________

Instructor ______________________ Section ______________________

1) In the space below draw an example of a linear translational symmetry.

2) In the space below draw a two-dimensional tilling using one of the shapes illustrated on page 100.

3) Name three objects that have at least one mirror plane of symmetry. Sketch the objects and discuss the symmetries present in each.

4) Draw examples of 3-fold, 4-fold, and 5-fold rotational symmetry.

5) Reproduce the objects drawn in Question 4 above, and indicate the positions of the mirror planes relative to the rotation axis of each.

6) For this question we will explore the symmetry of a cube using a standard die from a board game. Technically, any cubical object would work, but the die has each side uniquely marked, and allows for us to keep track of the various faces easily.

a) Hold the die between your first finger and your thumb. Place your finger on the center of the face with one dot, and your thumb underneath on the center of the six dot side. Rotate the die along the axis defined by your thumb and first finger. Name the faces you see. What type of rotational symmetry is this?

b) Place your first finger on the center of the edge between the one dot and two dot sides, and your thumb on the center of the edge between the five and six dot sides directly below. Hold the die so the three dot side is facing you. Rotate the die. Draw the four views you see, and describe this rotational symmetry.

c) Place your finger on the corner between the one, two and three dot sides and your thumb on the corner directly below between the four, five and six dot sides. Rotate the cube along this axis. Draw the six different orientations you can see. What is the nature of this rotational symmetry? How is it different than the rotational symmetries identified in parts a and b above?

Chapter 9 - Systems: Rates and Reservoirs

Introduction

The natural world is composed of many interesting and unique phenomena all of which are of great individual scientific interest. However, it is important for students of science to realize that nature is much more than the sum of its individual parts. In order to understand fully the natural world one needs to not only have a thorough knowledge of its innumerable components but also a complete understanding of the interconnections between the various parts. At the core of our study of the Earth as an integrated whole is the concept of natural systems. In this chapter we will explore this concept as well as the mathematical descriptions of the various interconnections that link such systems.

Systems

Nature is stunningly, and at times overwhelmingly, complex. That complexity is exhibited not only in the abundance and diversity of its individual components, but also in the astonishing variety of the interconnections between those components. How is it that scientists make sense of the abundance of natural complexity? How do they go about understanding the workings of the world? There are two important and highly interrelated processes of discovery that scientists use to address the complexity of nature.

Making sense of natural complexity.

The first process by which scientists study natural complexity is through exploring and describing the details. By cataloging and categorizing the diversity of nature it is possible to begin to establish meaningful classifications; that is, to place similar sets of details into understandable groupings. This kind of research is known as reductionism because the underlying activity is one of striving for understanding by breaking complex observations into smaller and smaller parts. By reducing the range of nature being studied, it is possible to make ever more detailed and exact measurements and observations of the phenomenon in question. The acquisition of detailed data enables scientists to more completely describe the very complexity that seems at times so overwhelming.

Reductionist science.

The second process by which scientists study natural complexity is through exploring and describing the interconnections between individual phenomena. By focusing on the relationships between individual parts, it is possible to develop a more complete understanding of all of the components of the system. For example, one might want to study all the chemical and physical properties of a suite of minerals. But true understanding of those minerals would await an exploration of how they are grouped together to form various types of rocks.

Process based science.

Definition of an Earth system.

When scientists begin the work of grouping and classifying natural phenomena they also develop ideas about how the components are linked together. The groupings of related natural phenomena and the processes by which they interact are generally known as systems. The most important difference between the study of nature through classifications and the study of natural systems is that classifications deal only with what is, while systems science deals with the ways in which the various components can and do interact. As such, a complete characterization of the processes of nature is as important as is the description of the myriad details. This integrated view of the Earth allows for the study of the wide range of large and small-scale interactions that control the climate, tectonics, evolution, and many other similarly complex phenomena.

Rates of Processes

Definition of a rate.

The scientific study of the Earth is an uniquely historical pursuit. Scientists are keenly interested in developing as complete a set of descriptions of the Earth's history as possible. However, in addition to this step-by-step historical journey, scientists are also interested in establishing how, and how fast, the Earth changed from one condition to another. In order to conduct such an analysis it is necessary for scientists to describe the changes they observe in terms of functional relationships. The calculation of the rate of a natural process is defined as the magnitude of some change divided by the duration over which that change occurred. Thus, rates are simply ratios of dependent variables relative to the independent variable time.

Graphical example of the calculation of a rate.

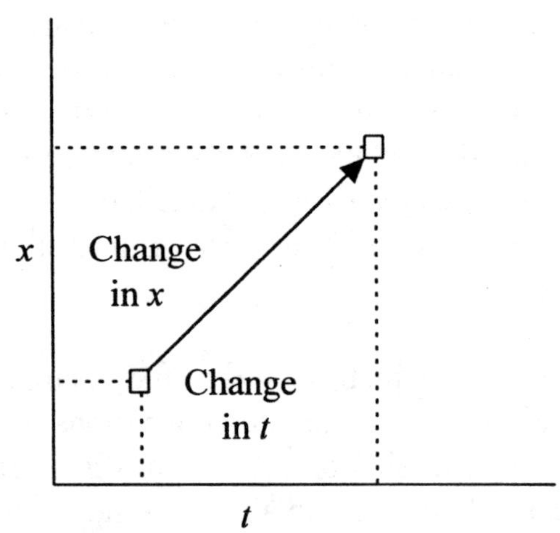

In the sciences and mathematics there are two important types of rates: average and instantaneous. The differences between these two are most easily understood in terms of the example provided by a drive across town at rush hour. How long does it take to get from one point to another, and what distance do you have to travel? Perhaps on a bad day it takes ½ hour to cover the 10 mile distance between your apartment and the mall. What was your rate of travel?

Calculation of an average rate.

$$10 \text{ miles} / 0.5 \text{ hours} = 20 \text{ miles per hour}$$

Thus, your average speed was 20 miles per hour. How does this value relate to the posted speed limits on the roads you travel? In some areas you are able to travel at rates of 25, 35, maybe even 45 miles per hour across town. Why then was your rate of travel so much lower than those potential speeds? Obviously, the reason is that the rate of speed we calculated for the entire duration of the trip reflects an *average* speed. At times you were traveling the posted limits of 25 or 45 miles per hour, but at other times you were forced to wait at traffic lights or slow in response to congested traffic. As you traveled it would have been possible to watch the speedometer and keep track of your rate of travel throughout the journey. That continuous measure of your rate of progress is known as instantaneous velocity. When driving down an open roadway your instantaneous rate was much faster than the long-term average rate, yet when stopped at a light your instantaneous speed was zero. The initial information given in this example (10 miles, ½ hour) provided no information about how the instantaneous rate of travel changed during the trip. That kind of uncertainty is very similar to the problems of interpolation described in Chapter 4, and is illustrated by this simple graph:

An example of average and instantaneous rates.

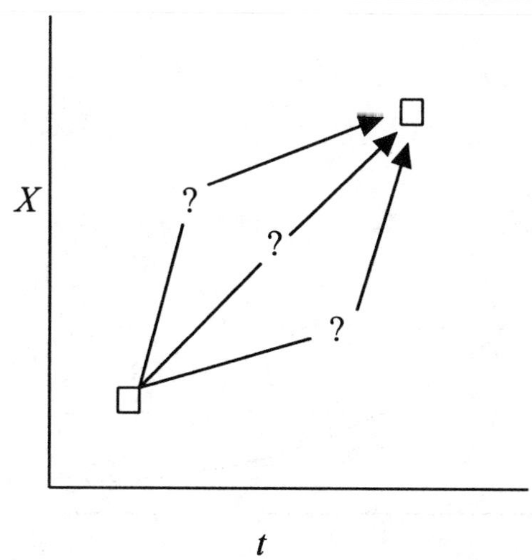

Graph illustrating uncertainty of instantaneous rate pathways between position/time observations.

Thus, it is clear that while average rate data are important in the Earth sciences, they provide little or no information of the position/time pathway actually taken during the completion of a geologic process

Developing a thorough understanding of the rates of natural processes is fundamental to understanding Earth systems. One of the most common stumbling blocks for students is distinguishing the difference between discussions of the changes of an increasing or decreasing functional relationship and the changes exhibited by the rate function of that process. Consider the following example:

Functional relationship between x and t.

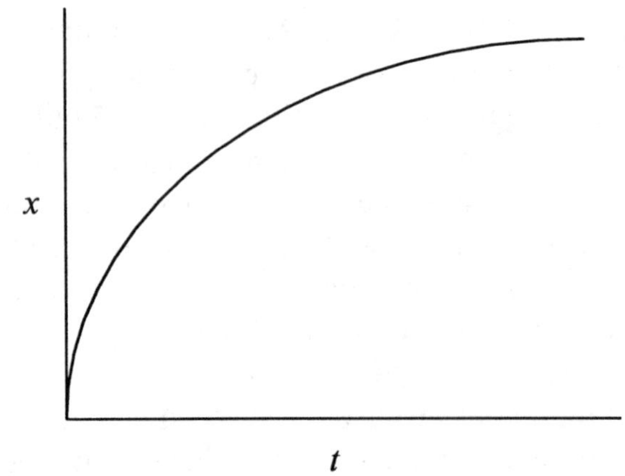

This graph illustrates a functional relationship between some variable x and time t. The shape of the curve is one of an increasing function, the slope of which becomes flatter as time passes. In effect, the observed change in the variable x is slowing as time passes. What does the rate function of such a graph look like?

Graph of the instantaneous rate of change of x with respect to t.

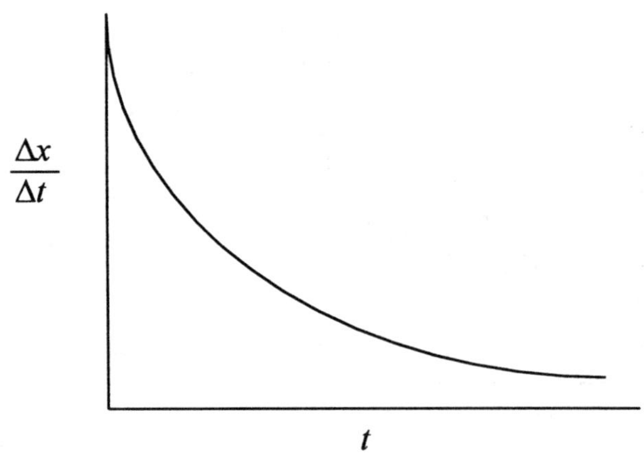

The decreasing rate function is clearly illustrated by the concave upward curve of this function. Notice that the rate of change of the variable x is symbolized by the ratio $\Delta x/\Delta t$, where Δ (pronounced delta) means "the change in" the variables x and t. Differentiation, one of the two major branches of the calculus, considers these types of rate functions in detail.

Natural systems are typically characterized by rate functions that look very similar to the example given. Why is that? What causes a natural process to slow as time progresses?

Why do rates slow with time?

Obviously, answers to those general questions are at least as complicated as the natural systems themselves. However, we can begin to understand why the decreasing rate function is so commonly encountered in natural systems by considering a few important facts.

From a purely physical standpoint, the presence of friction tends to slow the movement of materials as time passes. Slide a book across the floor. Initially the book is moving quickly, as time passes, its velocity slows due to the friction between the book's cover and the floor. Eventually the book comes to rest. The position and rate functions for such an experiment would look much like those illustrated on page 114.

Examples of causes behind the abundance of decreasing rate functions in natural systems.

Loss of energy to friction is an important control in the change of rate relationships exhibited by a very wide range of natural systems. However, our rush hour traffic example also serves as a successful model for many other types of natural systems. Processes studied by the Earth sciences are typically not continuous; rather, they are highly episodic. That is, activities occur within the natural system at some rate for some period of time only to be followed by long periods of inactivity.

Episodic character of natural systems.

Consider the case of sediment deposition during a flood. Initially a large amount of sand and silt accumulates as the flood waters recede. At what rate did those particles accumulate? Perhaps 1 meter of sediment was deposited during a flood that lasted one day. The rate of accumulation of sediment would then be calculated to be 1 m/day, a very rapid rate. However, the flood that produced the sediment was a very uncommon occurrence. As such, a long period of time passes before another flood accumulates any more sediment. This is the geologic equivalent to being forced to wait at a traffic light. Time passes, but no interesting activity occurs. Thus the *average rate* of many geological processes decrease the longer the observation lasts. These intervals of time between episodic events can be thought of as "waiting around" time — time that is not recorded by the deposition of sediment, but that must be included in the calculation of average rate. Thus, the combination of friction, and episodic activity within natural systems leads to the development of decreasing rate functions for nearly all natural phenomena.

The concept of "waiting around" time in Earth science.

Reservoirs and Fluxes

Earth systems are dynamic. The atmosphere, the oceans, and the land's surface are always in a state of change. Some changes are very rapid and catastrophic, others are very slow and gradual — almost unnoticed by human observation. In order to understand these dynamic systems, it is necessary to consider both the materials and the processes that connect them.

The water cycle as an example of a major Earth system.

Fluxes and Reservoirs defined.

One of the most easily understood of the major Earth systems is the water cycle. This system, like all the others that operate on Earth, can be broken into two major structural components. There are reservoirs and fluxes. Within the water cycle, the reservoirs are bodies of water, while the fluxes are the processes of transfer between those bodies. The oceans represent the largest reservoir of water on Earth, the water vapor in the atmosphere one of the smallest. What are the fluxes between those two reservoirs? The movement of water from the surface of the sea to the atmosphere is known as evaporation and can be expressed in terms of some volume of water transferred over some unit of time. The return of water from the atmosphere to the ocean we of course term precipitation (or rainfall). Thus, fluxes are simply rates of transfer between bodies of water.

Ocean/atmosphere fluxes.

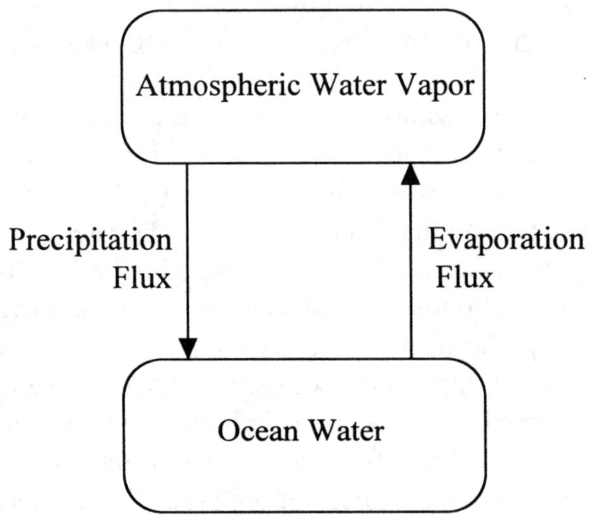

However, the Earth system that deals with the water cycle is much more complex than just two reservoirs and two fluxes. There are surface water reservoirs (lakes and rivers) as well as ground water and biological reservoirs. Of course there are then many other fluxes including those between surface water and the sea illustrated by the discharge of rivers and the infiltration of surface water into the ground or the evaporation off the surface of the Earth. Likewise, springs return water from the ground to the surface water system.

Complete characterization of a dynamic Earth system requires both the identification and the measurement of all the reservoirs as well as all the fluxes between them. Failure to identify or accurately describe either a flux or a reservoir leads to an incomplete understanding of the Earth system being studied. Ongoing advances in science allow us a more accurate and complete understanding of the dynamics of complex systems. However, the more we learn, the more we find connections and dependencies that we had failed to adequately account for in past models. Refining our understanding of Earth systems is a continual process of scientific discovery.

Earth systems science is an ongoing subject of study.

System Dynamics

Within Earth systems fluxes typically move materials from one reservoir to another. In the example of the ocean and the atmosphere evaporation acts as a flux that removes water from the surface of the sea and returns it to the atmosphere in the form of water vapor. This is a constant and ongoing process occurring worldwide. However, the oceans do not dry up. Why does this constant flux of water not eventually deplete the oceans of their volume? The obvious reason is that there is a second flux, known as precipitation, that returns water from the atmosphere to the sea.

In a simple ocean/atmosphere world, the exchange of water between the sea and the air would eventually reach a balance. The rate of evaporation would equal the rate of precipitation over the long-term. That is, the oceans would not increase or decrease in volume; likewise the amount of water vapor stored in the atmosphere would remain constant. This situation is known as a steady-state. The rates of two competing fluxes balance, to produce constant volumes of water in each reservoir, and is illustrated by this graph of volume of water in the oceans:

Definition of steady-state systems.

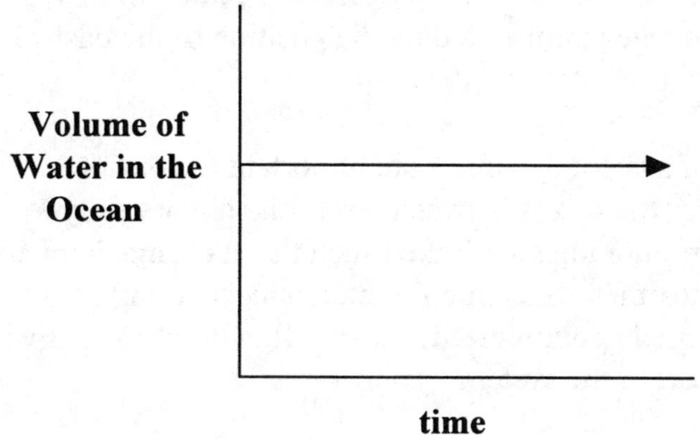

Graph of a steady-state.

Mathematical equation of a steady-state.

In mathematical terms, the relationship described above can be written as:

$$\text{Volume}_{\text{ocean}}(t) = \text{Volume}_{\text{ocean}}(t\text{-}1) + \text{Flux}_{\text{precipitation}} - \text{Flux}_{\text{evaporation}}$$

That is, the volume of water in the oceans at any time t is equal to the volume at some previous time t-1 plus the difference between the precipitation and evaporation fluxes.

Definition of a system in dynamic equilibrium.

As long as the fluxes of evaporation and precipitation are exactly equal, the volume of the ocean must to be constant. However, most Earth systems are not characterized by perfect constancy in their fluxes. Rather, the rates of exchange between reservoirs are typically found to vary around some long-term average. This situation is one where in general the fluxes are equal, but for short periods of time they are slightly out of balance. Systems that exhibit such behavior are said to be in dynamic equilibrium, an example of which is illustrated by this graph:

Graph of a dynamic equilibrium.

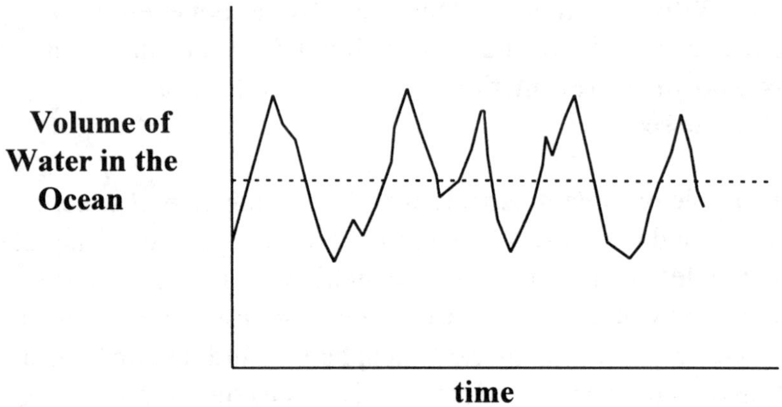

In this case, the volume of water in the ocean is found to vary as the rate of the precipitation flux exceeds or falls short of the evaporation flux for short periods of time. Yet over time, the system is found to have a general equilibrium value around which the variation occurs. In this example the long-term average is illustrated by the position of the dashed line.

Two other types of system dynamics are important to consider. Typically in Earth systems, the fluxes between reservoirs are not fully independent of each other, but rather are linked such that a change in one flux causes a change in the other. Such highly interconnected Earth systems can become extremely complicated and are often highly sensitive to small changes in any part of the system.

Consider the linkages between evaporation and precipitation. Why would our water-world example be expected to reach a long-term dynamic equilibrium? The answer can be found in the interconnection between the two fluxes. Precipitation occurs when water vapor condenses in clouds and falls to Earth. When a cloud is overhead, part of the sun's light energy is blocked by the cloud. As such, the surface of the sea under a cloud would be expected to be cooler than the surrounding water that is experiencing clear skies. Thus, the cooler seawater would evaporate at a slower rate than other regions of the sea, and the net evaporation flux would be reduced as a result.

Geological example of a negative feed-back system.

Earth systems such as the one described above are said to be experiencing negative feedback. Negative feedback results in the gradual reduction in the magnitudes of variation of fluxes and thus a return towards the long-term equilibrium value for the reservoir.

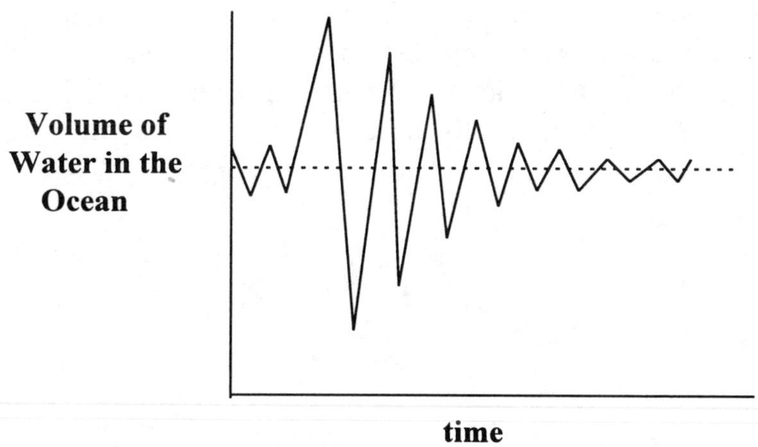

Graph of a negative feed-back system.

The second important type of dynamics experienced by Earth systems is known as positive feedback. Everyone has at one time or another experienced positive feedback. If a microphone is held too close to a speaker, the sound from the speaker is amplified through the microphone, and returns as sound again from the speaker. This looping of sound from speaker to microphone to speaker quickly results in an ear piercing screech.

Within Earth systems, potential positive feedback loops exist within the climate system. As climate cools, precipitation from the atmosphere takes on the form of snow. Of course, snow blankets the ground with a layer of whiteness. The energy from the sun, which serves to warm the Earth is reflected more efficiently off white snow than it is off soil or rock. Thus, a snow covered Earth starts to cool more and more as a larger percentage of solar energy is reflected back to space. If

Geological example of a positive feed-back system.

allowed to continue long enough, the planet could become locked into a permanently frozen state.

Graph of a positive feedback system.

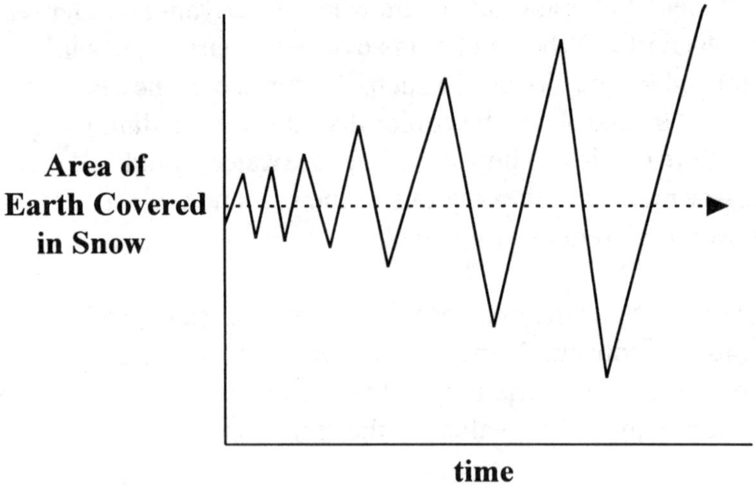

Positive feedback loops such as this are uncommon in nature, because if they were to experience any disturbance they would rapidly diverge from equilibrium and in time reach some extreme value. That is, such systems are strongly dynamically unstable.

Chapter 9 - Exercises

Name ______________________________ Date ______________________________

Instructor _________________________ Section ____________________________

1) A rock falls of the side off a mountain slope and moves a total horizontal distance of 25 meters in 5 seconds. Illustrate this situation with a simple sketch. How would you express this rate of movement? Is the rate you calculated an average or instantaneous rate? How would the two differ in this particular situation?

2) A particle of silt, carried in suspension in a mountain river, moves a distance of 50 meters in 25 seconds. How would you express this rate? Is this a measure of an average or instantaneous rate? How and why would the instantaneous rate of motion of the silt grain differ from the rate you calculated?

3) You are studying a closed sedimentary basin in which sediments eroded from the adjacent mountains are accumulating. Between two well-dated horizons is an important feature — perhaps an unusual fossil — and you want to know its age. If the sediments accumulated continuously at a constant rate, describe how you would date the feature. If the sediments were found to have accumulated discontinuously, describe the error you would make if you assumed that accumulation was continuous. Use sketches to illustrate your reasoning.

4) Consider a mineral grain exposed at the Earth's surface that is being subjected to chemical weathering. The rate of chemical alteration of the grain is found to decrease with increasing time. Sketch graphs illustrating the total chemical alteration experienced by the grain through time, and the decreasing rate function. Label the axes appropriately.

5) A glacier can be thought of as a simple reservoir of ice. There are two important fluxes into and out of glaciers. An accumulation flux brings new snow to the upper reaches of the glacier which through time is transformed to ice. A second flux, known as ablation, is the sum of melting and direct evaporation (sublimation) off the ice. Sketch a simple box model of a glacier ice reservoir and these two fluxes.

6) If the volume of ice in a particular glacier is known to be $4.35 \times 10^6 \text{ m}^3$, the accumulation flux during the following year is $2.53 \times 10^5 \text{ m}^3/\text{year}$ and the total ablation flux is $1.94 \times 10^5 \text{ m}^3/\text{year}$, calculate the change in the reservoir of ice. Show all work.

7) If these rates of accumulation and ablation were to continue over a long period of time — perhaps several hundreds of years — sketch a graph of the change in the reservoir size of the glacial ice over time. Label all axes appropriately.

8) Consider the concepts of steady-state, dynamic equilibrium, negative feed-back, and positive feed-back systems. Provide natural examples of each, and sketch graphs of the various systems over time.

Chapter 10 - Fractals: Scale and Pattern

Introduction

Nature is typically not composed of simple lines and planes nor of the smooth surfaces of squares and spheres. Rather, when observed in detail, the edges of nature are found to be somewhat fuzzy and considerably more rough and rugged than the sharp points and crisp lines of elementary geometry. This ruggedness, this apparent irregularity, is a hallmark of natural systems. Classical plane and solid geometry has proven to be a very powerful descriptor of most aspects of nature. The success of this geometry is attested by its antiquity. However, over the last quarter century scientists have become increasingly aware of and interested in the detailed ruggedness of nature. Fractal geometry serves as a means to describe that fuzzy complexity of nature as well as provide a foundation for the understanding of the processes which give rise to many of those irregular and complex forms.

History and Development

The word 'fractal' is derived from the Latin word *fractus* which mean broken or irregular, and was coined by the founder of this branch of mathematics, Benoit Mandelbrot, in the1975 book *The Fractal Geometry of Nature*. Fractal geometry is a mathematical description of the irregularity of nature. Traditional geometry describes exceedingly regular, and typically highly symmetrical, shapes. However, it has long been recognized that while serving as a reasonable first approximation for many aspects of nature, the simple geometry of lines and planes — of surfaces and volumes — was decidedly inadequate for the description of both the details of nature, as well as many of its largest shapes and forms. Yet fractal geometry has proven to be more than just a new and improved descriptor of nature's complexity. The mathematics of fractal geometry also serves as the basis for much of the analysis of processes that give rise to these irregular natural shapes. As such, any complete discussion of the mathematical foundations of the Earth sciences must include some consideration of nature's fractal form.

Definition of fractal geometry.

Dimensions

Since the time of the ancient Greeks, the geometric description of nature has been based upon an approximation of regularity. That approximation has involved analysis of the simple shapes described by points, lines, planes, and volumes. Each of those various geometric forms is defined in terms of its dimension — the measure of position in one direction. Thus volumes are three-dimensional shapes with width, depth, and height, surfaces such as planes are two-dimensional, while lines are expressed in terms of a single dimension. Interestingly, points are defined to occupy no space and as such are dimensionless. These three

Definition of a dimension.

Approximating irregularity with elementary geometric forms.

dimensions that define the geometries of perfect shapes are very useful as descriptors of nature. The smooth surface of water in a pond can be described as a plane of two dimensions, just as the sharp edge between two crystal faces describes a line. As such, there are many aspects of nature for which the three dimensions are adequate approximations; however, most of nature is composed of less regular shapes.

How do scientists describe these many irregularities of nature? One of the first and most direct methods is to use the various regular geometries to approximate more complex lines, surfaces or volumes. Essentially this requires an approach to geometric description that is very similar to the notion of interpolation discussed in Chapter 4. By taking short segments of lines, or simple areas or volumes, it is possible to begin to approximate even very complex objects. While this is a good approach, it still does not directly describe the shape in question. Often, scientists have been forced to resort to such ill-defined terms as wiggly or squiggly, relatively smooth or irregular, lumpy or bumpy. While carrying a general visual connotation, these colorful words fail to provide a solid description of the irregularity of nature. How irregular is a line, and how would we compare its irregularity to that of some other line?

The beauty and power of fractal geometry is that it provides a quantitative description of the irregularity of nature. Fractals allow for mathematical comparisons and descriptions through which we can begin to study the underlying causative processes which form and shape the irregularities of nature.

Fractional Dimensions

How wiggly is a line, how bumpy is a surface? Fractal geometry has allowed for the quantification of those irregularities, and serves as a basis for comparison between the various shapes exhibited by natural systems. A geometric line — perfect in its linearity, extending infinitely in two directions — defines a single dimension. Thus, lines are said to be objects of dimension 1.

A line has the geometric dimension 1.

Line of Dimension 1

This line has infinite length, and infinitely small width and height. Its size can only be expressed in terms of the direction in which it is trending. Imagine taking that simple, continuous, straight, perfect line and placing a small irregularity somewhere along its length. Such a bump or wiggle in the line breaks its simplicity. Indeed, it radically alters its shape.

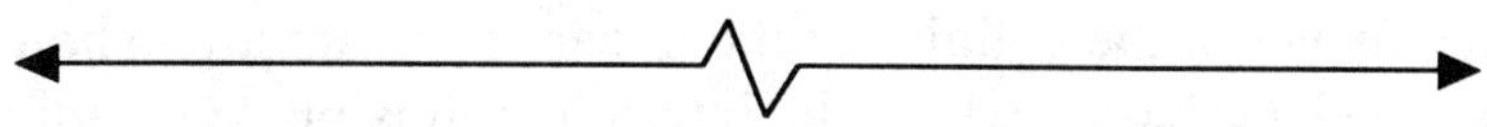

A Line With an Irregularity

A perfect line disrupted by an irregularity.

How does one now describe this line? It is no longer geometrically perfect. Whereas previously it had only length, it now can be described as having both a length as illustrated by the directions of the arrows and, in this case, some degree of height as illustrated by the amplitude of the irregularity. The line is no longer precisely of dimension 1; yet it is not fully dimension 2. Fractal geometry describes such a line as having a *fractional dimension* somewhere between a perfect line (1) and a perfect plane (2). One of the most interesting and unusual aspects of fractal geometry is that it allows shapes to be described in terms of dimensions that need not be integers. Thus, Mandelbrot defined a *fractal dimension* to be the spatial dimension of an object that may take on either a fractional or an integer value.

Irregular objects can have fractional dimensions.

Definition of a fractal dimension.

Irregular lines with fractal dimensions between 1 and 2 are said to partially fill the plane, and the degree to which they do so is represented by the magnitude of the fraction. Likewise, irregular surfaces are said to be partially volume-filling as their irregularity grows from a perfect geometric plane to a surface of great complexity. The concept of a fractional dimension allows us to describe and compare the complex irregularities of nature, but how is it that the dimension of any particular irregular object found?

Lengths of Coasts and Frontiers

The first direct application of the foundational concepts of fractal geometry was applied to the description of the lengths of irregular lines, such as those observed for coastlines and political boundaries or frontiers. Unquestionably, such geographic lines are irregular. There are many twists and turns reflecting the sum of the effects of the composition of bedrock, the tectonic forces which have acted on the region, and the climatic condition under which the land has been sculpted. How does one describe this irregularity?

In the late 1960's Mandelbrot rediscovered the work of the British scholar Lewis Fry Richardson. Richardson had found that the measured length of an irregular line increased systematically as a function of the decreasing length of the measuring unit. What this means is that if the length of an irregular line is approximated by adding up a series of straight-line measurements the total length increases as the length of the line segments decrease. That observation is and of itself not overly surprising. It is expected that the approximation would become more accurate as the precision of the measurement increases. That is to say, as learned in Chapter 4, the validity of the linear approximation should increase as the length over which the interpolation is made is reduced. However, the fascinating point is that the measured length does not converge towards a finite value. Rather, the length continues to increase as the length of the measuring stick decreases. Even more importantly, it does so according to a very systematic mathematical law.

Example of extreme irregularity in the northwest coastline of Scotland.

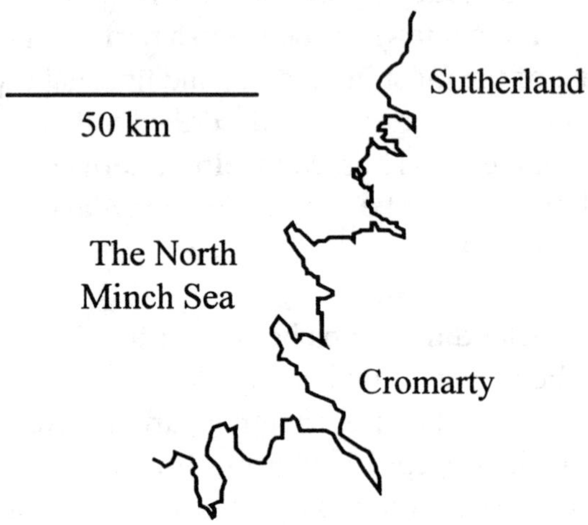

Graphical representation of a Richardson analysis.

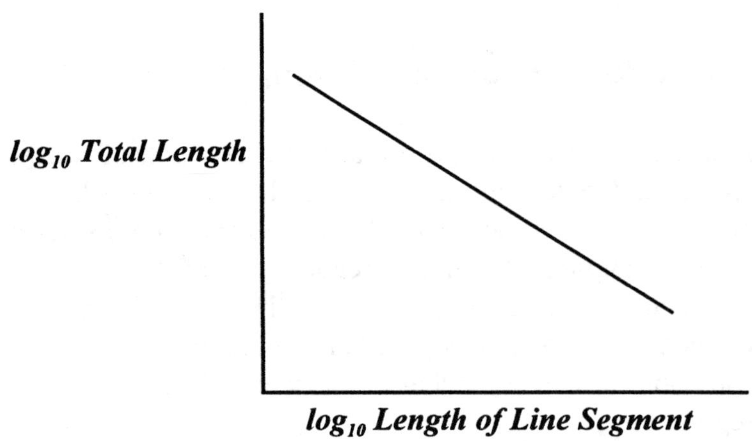

This mathematical relationship between the size of the approximation interpolation distance and the total measured length is expressed by the formula:

$$L \sim= ls^{(1-D)}$$

Elementary formulation of the fractal dimension.

Where L is the total length of the feature, *ls* is the length of the line segment, and D is the fractal dimension of the irregular line.

Scale Invariance

Definition of scale invariance.

What does it mean to say that the length of the coastline is found to increase without limit as the length of the measurement interval decreases? It runs counter to our commonly held concepts of reality to find that an infinite length can be contained within a finite area. However, in many cases this is the exact, and undeniable, fact. Systems that are found to take on such unbounded geometries are said to be scale invariant. That is, they are independent of the scale of observation. In this case the complex curvature of the northwest coast of Scotland would look generally similar if observed as illustrated on page 130 or at much higher resolution. Obviously the details of the curvature would be different, but the important point is that the degree of irregularity, as expressed by the calculated fractal dimension D, would be found to be invariant. Such independence of scale is extremely common in the Earth sciences, and is the reason why scale bars or other indicators of absolute size must always be included in photographs of geometrically interesting features.

Natural scale invariance tends to be bounded.

Nature is assuredly complex and that complexity often transcends scale. However, not all complex systems are scale invariant over all possible magnitudes. Natural systems are bounded, and while scale invariance does commonly extend over many orders of magnitude of observational lengths, it does not typically extend infinitely to smaller or larger scales. Yet the presence of invariance over large ranges of observational scale is an important aspect of nature, one that fractal geometry can now describe, and one that scientists can now more adequately attempt to explain.

The study of fractal geometry is nearly as complex as nature itself. New and exciting applications of these concepts are being discovered and applied to the Earth sciences every day. This brief discussion has been intended to remove a bit of the uncertainty which surrounds the subject, and to provide an introduction to this new and expanding mathematical technique.

Chapter 10 - Exercises

Name ______________________ Date ______________________

Instructor ______________________ Section ______________________

1) List and describe three geometrically complex objects you have learned about in your Earth science class this term. Discuss briefly the nature of their complexity, and measurements you would make to describe that complexity.

a)

b)

c)

2) Consider the construction of the following complex line. Start with a line segment of length 1.

Subdivide that line segment into 3 equal lengths and then replace the middle length with a square-shaped irregularity of side length 1/3.

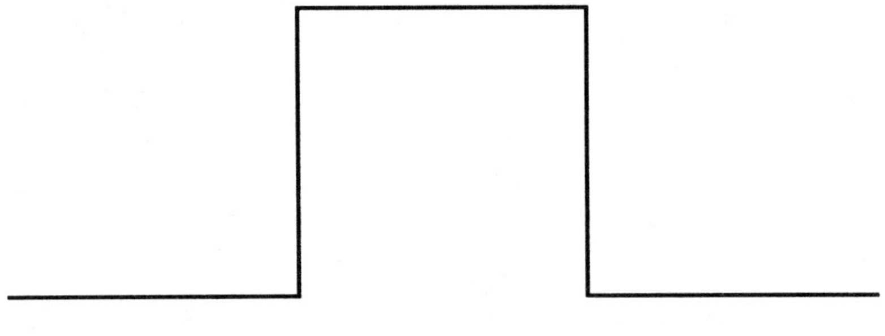

The total length of the line has now increased to 5/3 from its original length of 1. And the size of each smaller segment is now 1/3 the original length. As a next step take each of the five new straight line segments and subdivide them into three equal lengths, inserting a square-edged irregularity at the location of each middle segment. Draw the effect of such a further complication directly on the illustration above.

What is the total length of the line now? What is the size of the individual straight-line segments?

Repeated application of this mechanism of inducing irregularity causes the total length of the line to increase according to the equation:

$$L(n) = 5^n$$

where L is the total length, and n is the number of times you increase the complexity by inserting more square boxes in each line segment. Interestingly, the length of the individual line segments are found to decrease according to the equation:

$$ls(n) = 1 / (3^n)$$

where ls is the length of the line segment at any step n. In the space below, sketch the result of applying this process one more time to the pattern you drew onto the illustration on page 134.

3) Using the equations above, calculate the changes in total length and length of line segments for the first ten iterations of the steps described above. Calculate the base 10 logarithm of each value. Fill in the table below with the values you calculate.

n	L(n)	*ls*(n)	log(L(n))	log(*ls*(n))
0				
1				
2				
3				
4				
5				
6				
7				
8				
9				
10				

4) On the following page construct a graph similar to the one illustrated on page 130, with $\log_{10} ls(n)$ as the horizontal axis, and $\log_{10} L(n)$ as the vertical axis. Let the horizontal axis scale range from -5 to 0, and the vertical axis scale range from 0 to 7. Describe the relationship illustrated by the graphed points in the space below.

5) Calculate the slope of the line using the formulation described in Chapter 6. Show all work in the space below.

6) Finally, using the notion that the fractal dimension of an irregular curve is given by the equation $D = - m$, where *m* is the slope of the line calculated above, calculate the fractal dimension of the complex line. Comment upon the meaning of this value. Can you find another way to *directly* calculate the fractal dimension?

Appendix I

Conversions of Length

1 cm = 0.3937 in
1 m = 100 cm
1 km = 1000 m
1 km = 0.6215 miles
1 light-year = 9.461×10^{15} m

Conversions of Area

$1\ m^2 = 10^4\ cm^2$
$1\ m^2 = 10.76\ ft^2$
$1\ cm^2 = 0.155\ in^2$

Conversions of Volume

$1\ m^3 = 10^6\ cm^3$
$1\ cm^3 = 1$ ml
$1\ l = 1 \times 10^3\ cm^3$
1 l = 0.264 gal
$1 m^3 = 1 \times 10^3$ l

Conversions of Mass and Force

1 kg = 1000 g
1 kg = 2.20 lbs at sealevel

Conversions of Time

1 year = 3.16×10^7 s
1 day = 8.64×10^4 s

Conversions of Temperature

$T_k = T_c + 273.15$
$T_c = 5/9\ (T_f - 32)$